gRiddLeRS
Logic Puzzles

Triddlers
Volume 1

Smart Things Begin With Griddlers.net

Griddlers Logic Puzzles: Triddlers Vol. 1

Published by: Griddlers.net
a division of A.A.H.R. Offset Maor Ltd

Author: Griddlers Team
Compiler: Rastislav Rehák
Cover design: Elad Maor
Contributors: amsterdam, ANA_LILI, Aranlyde, argosgold, ariari1660, attackerad, bedirhan, bilbobilbo, cekpompiere, Charlotta, Don_Roberto, efe929, elad, Elena_13, ferloft101, galash, galocha2003, gneale, hentron, Heracleum, hertseltsur, hi19hi19, jadespade, jiunndar, jkxdea, kayton, ledka, lenchik, Lenchik, lloydirving, lovinbayb4e, luweewu, madziasek, makie, malymichu, mariamnur, MathLogicLady, meszi, Miepsah, MissAdler, music_love84, mustafademirbas, myreg, norwayLiz, painter100, petersong, podjedzona, Ra100, sandyeggan, shadow2097, Sliam, steph22, Teeaz, Terppa, tortois3, twig007x, unwound, Vargflickan, wieralee, zjmonty.

ISBN: 978-9657679302

More information:
Email – team@griddlers.net
Website – http://www.griddlers.net

Triddlers Rules and Examples

Triddlers are picture logic puzzles that use number clues around a grid to create an image.

The clues encircle the entire grid in three directions.

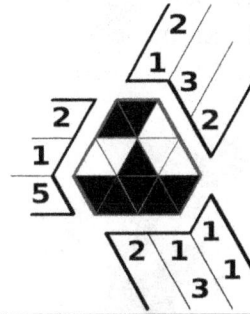

Each clue indicates a group of contiguous triangles of like color.

Between each group there is at least one empty triangle.

The clues are already in the correct sequence.

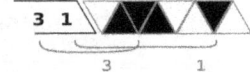

Groups of different colors may or may not have empty triangles between them.

In Black and White puzzles the clues are always black and the background is always white.

In Color puzzles the background may or may not be white.

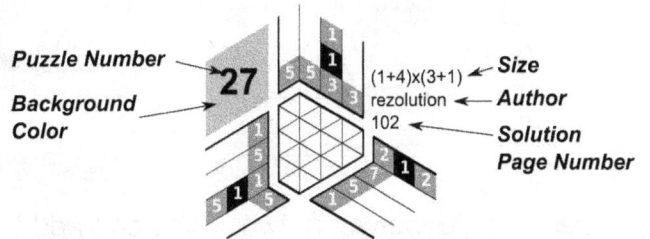

The colored shape on the puzzle top-left corner indicates the color of the background.

Solving a Puzzle

The gray arrow marks the direction of the clues as well as the direction they should be placed on the grid.

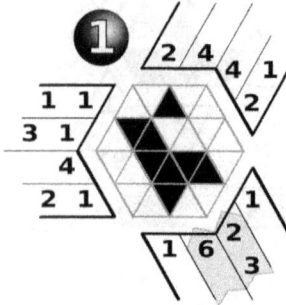

If we perform overlapping counting and count from the bottom up and from the top down, we can place 5 triangles of clue **6** on the grid. We can do the same for clues (**2,3**).

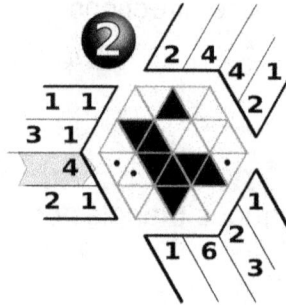

Clue **4** already has 4 triangles on the grid. We can mark the rest as background color.

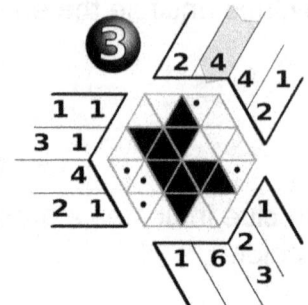

There are already 1 triangle and 2 triangles in the line of clue **4**.
We will add a triangle between them to make it a group of 4.

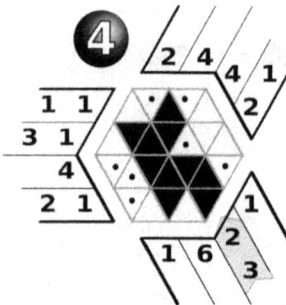

The triangle between two blocks has to be marked as background. Now we can complete the line of clues (**2,3**).

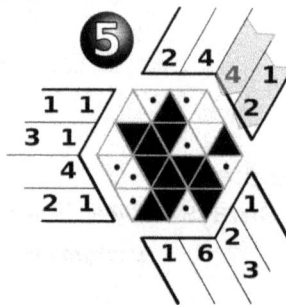

We can add the 4th triangle needed for clue 4 and complete the line of clues (**1,2**).

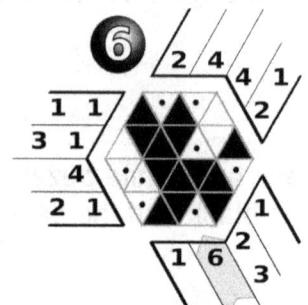

There is only one empty triangle left to complete clues (**1,1**) and finish the puzzle.

1

(17+8)x(6+22)
ANA_LILI
98

2

(5+5)x(5+5)
zjmonty
98

3

(10+10)x(10+10)
zjmonty
98

4

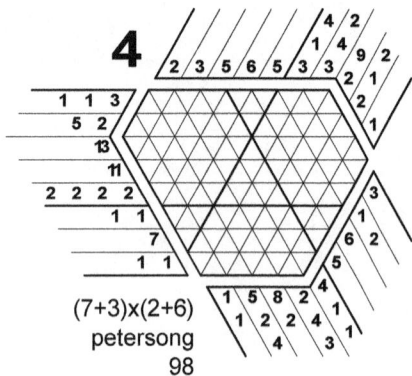

(7+3)x(2+6)
petersong
98

5

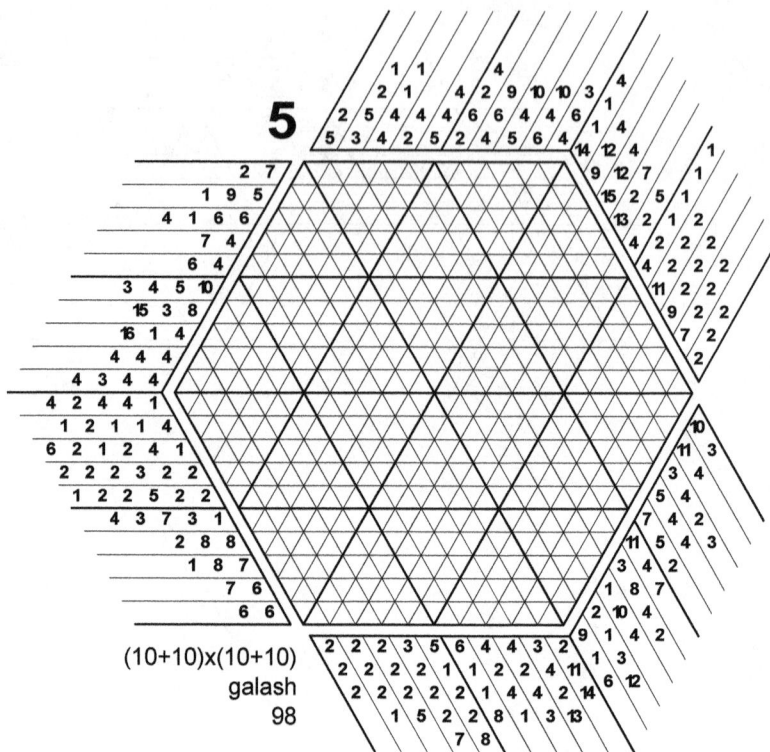

(10+10)x(10+10)
galash
98

6

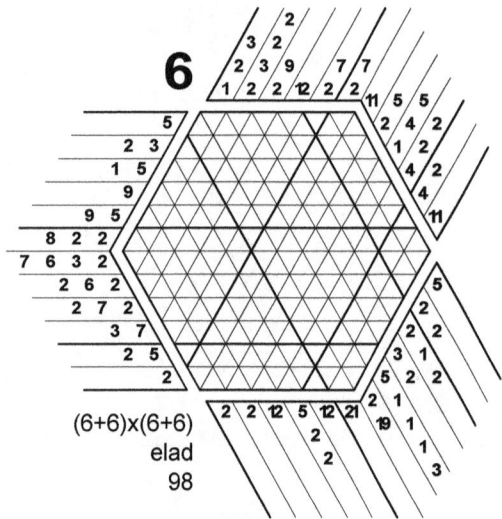

(6+6)x(6+6)
elad
98

7

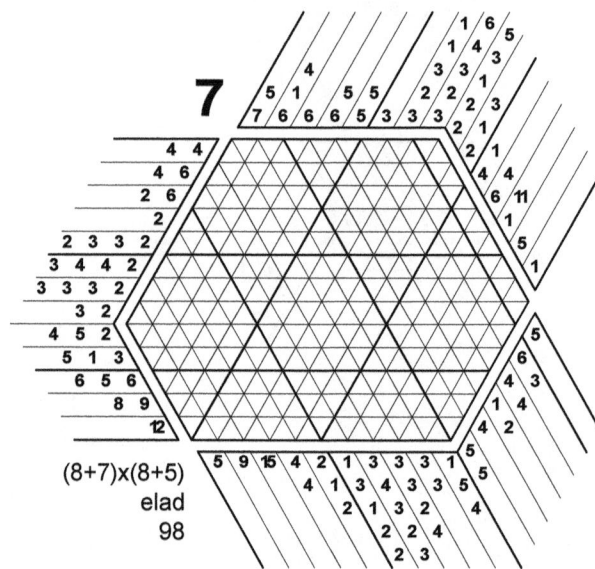

(8+7)x(8+5)
elad
98

8

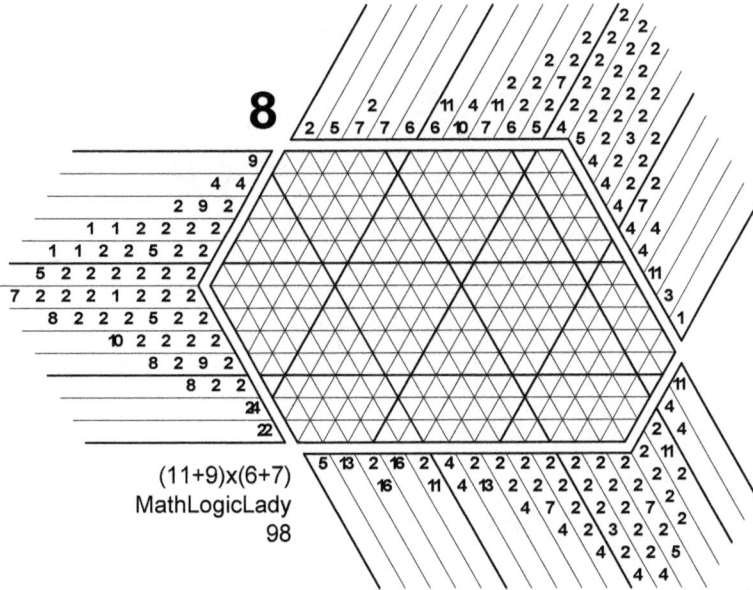

(11+9)x(6+7)
MathLogicLady
98

9

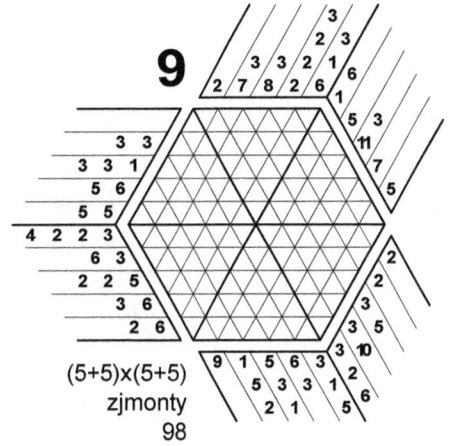

(5+5)x(5+5)
zjmonty
98

10

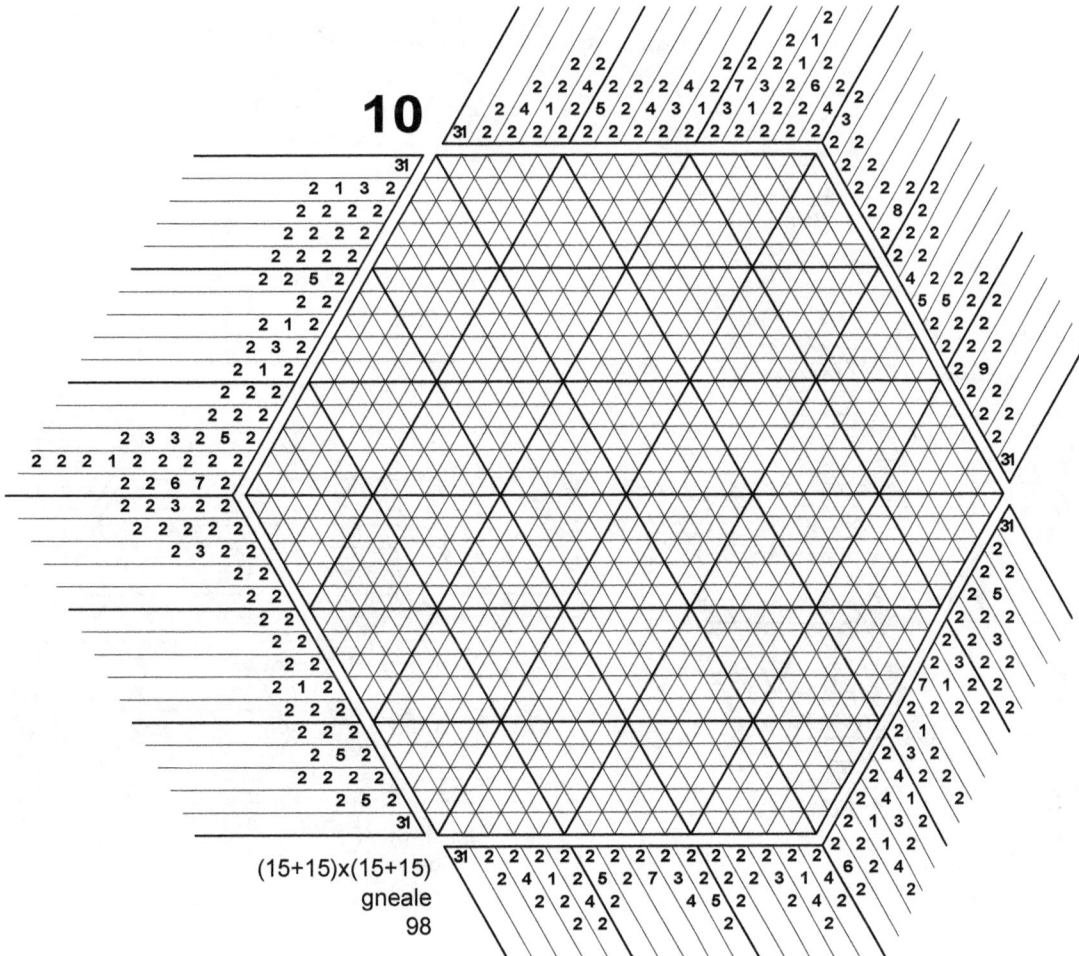

(15+15)x(15+15)
gneale
98

11

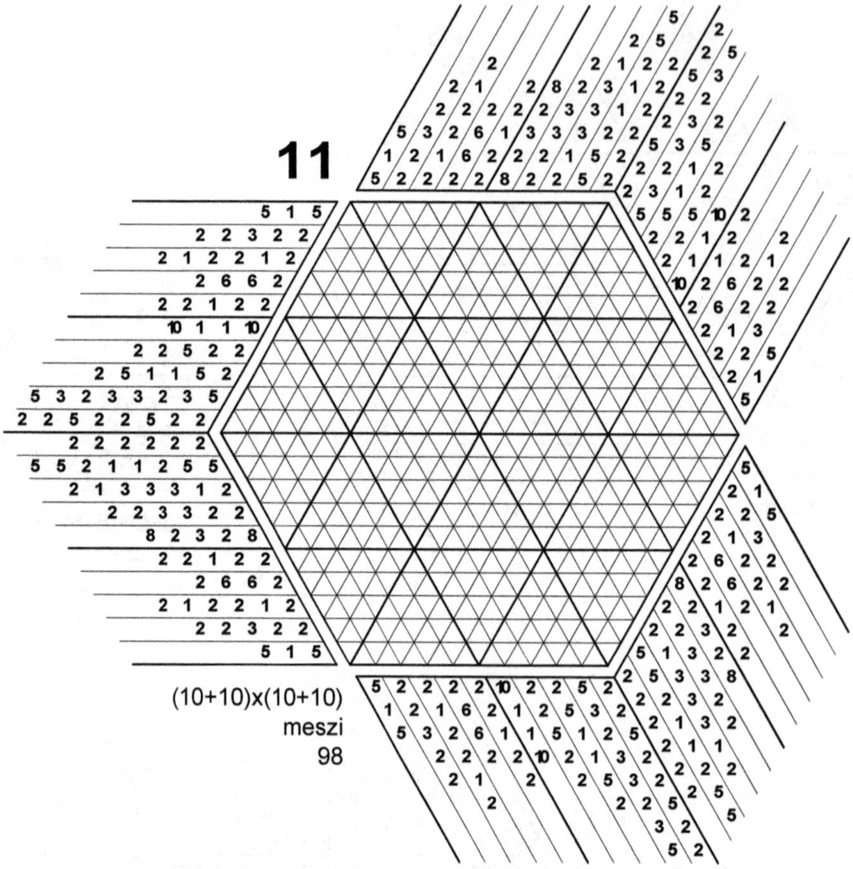

(10+10)x(10+10)
meszi
98

12

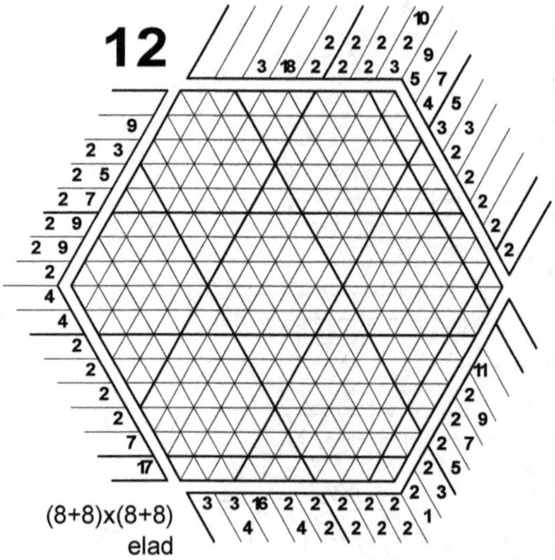

(8+8)x(8+8)
elad
98

13

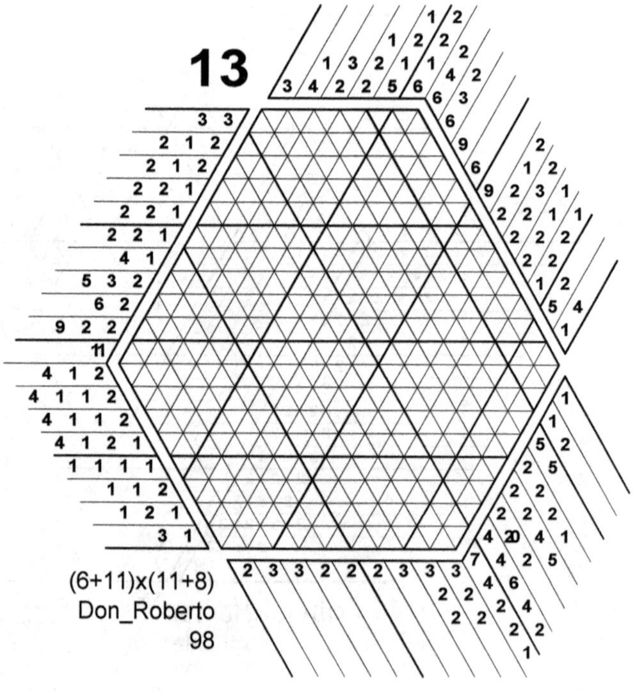

(6+11)x(11+8)
Don_Roberto
98

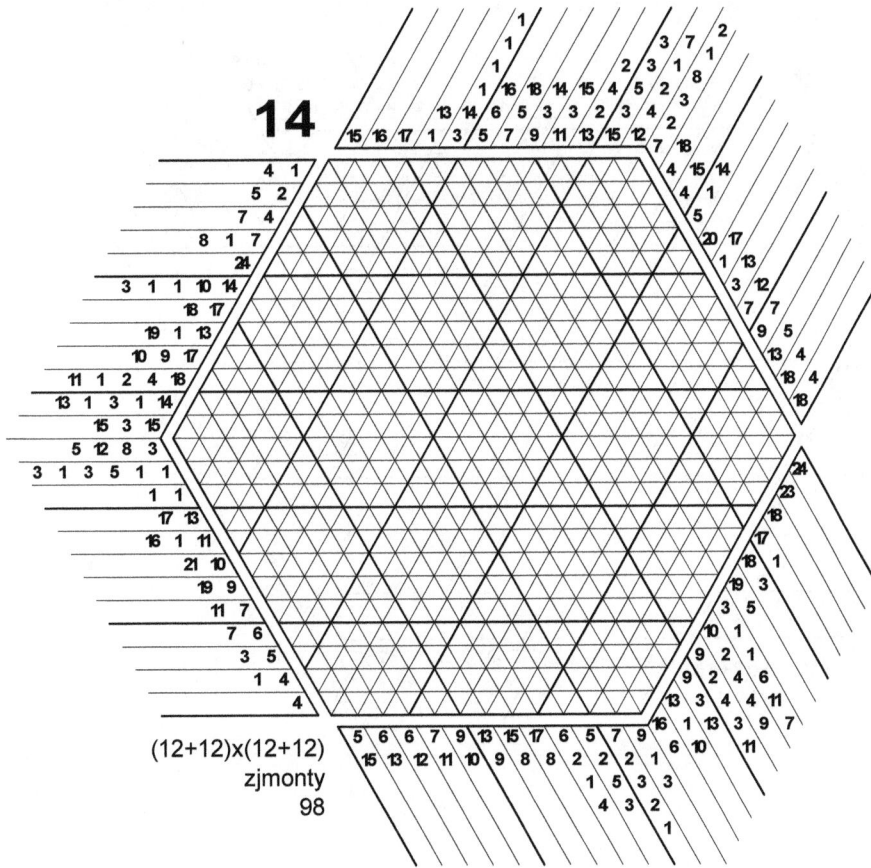

14

(12+12)x(12+12)
zjmonty
98

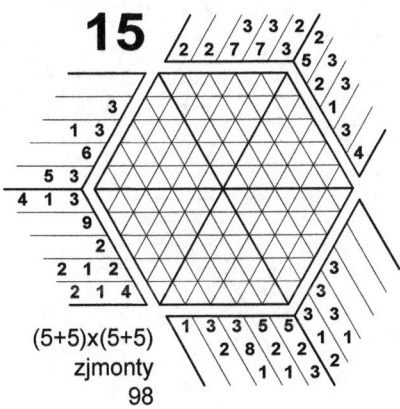

15

(5+5)x(5+5)
zjmonty
98

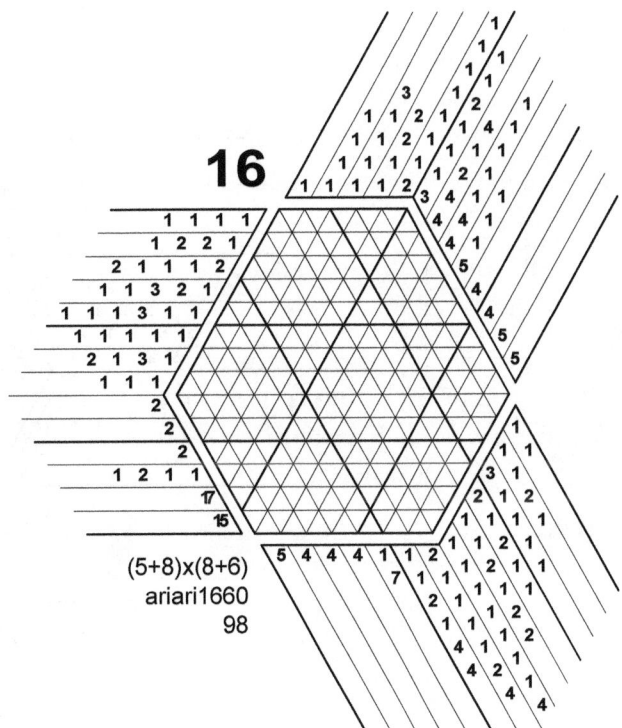

16

(5+8)x(8+6)
ariari1660
98

17

(21+7)x(7+15)
ANA_LILI
98

18

(10+10)x(10+10)
zjmonty
98

19

(10+10)x(10+10)
Terppa
98

20

(14+14)x(12+15)
twig007x
98

21

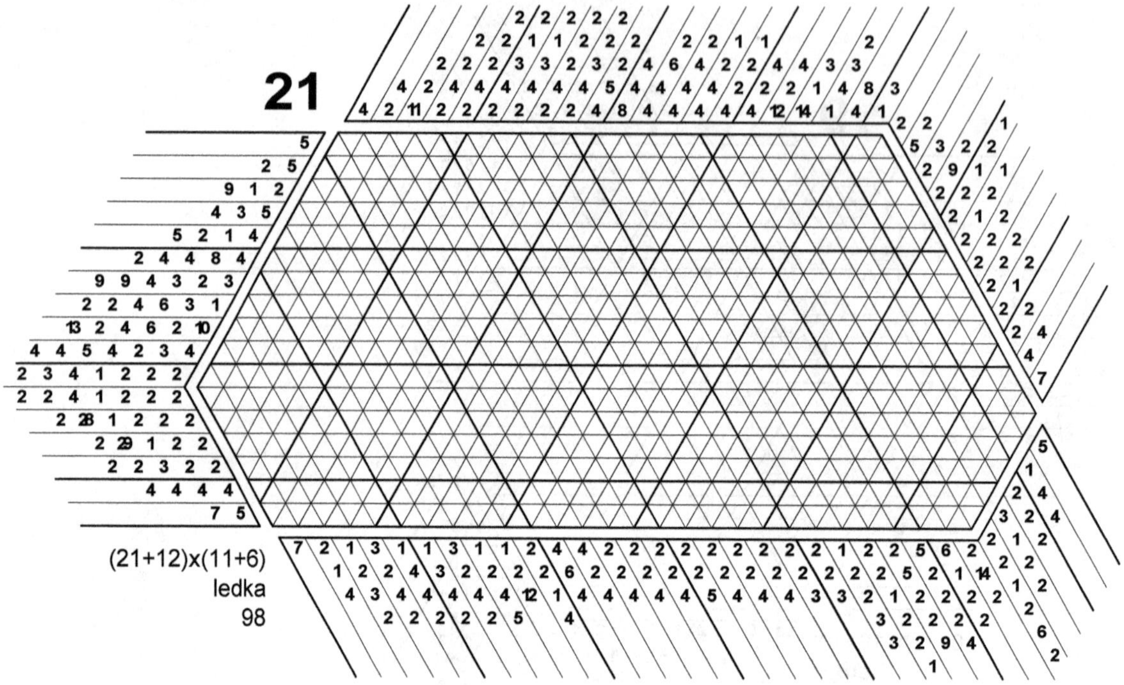

(21+12)x(11+6)
ledka
98

22

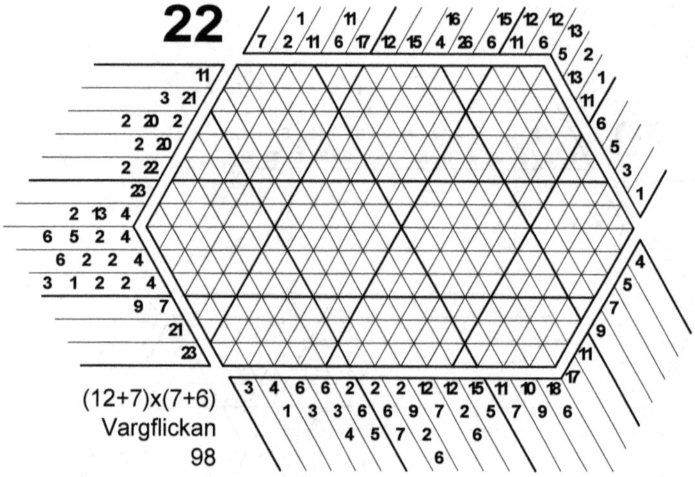

(12+7)x(7+6)
Vargflickan
98

23

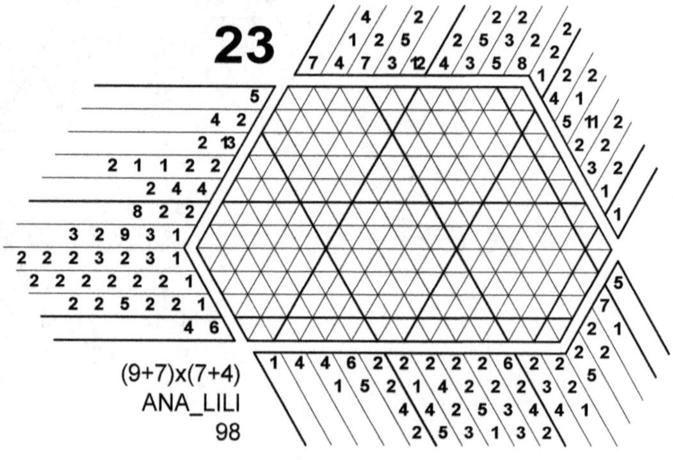

(9+7)x(7+4)
ANA_LILI
98

24

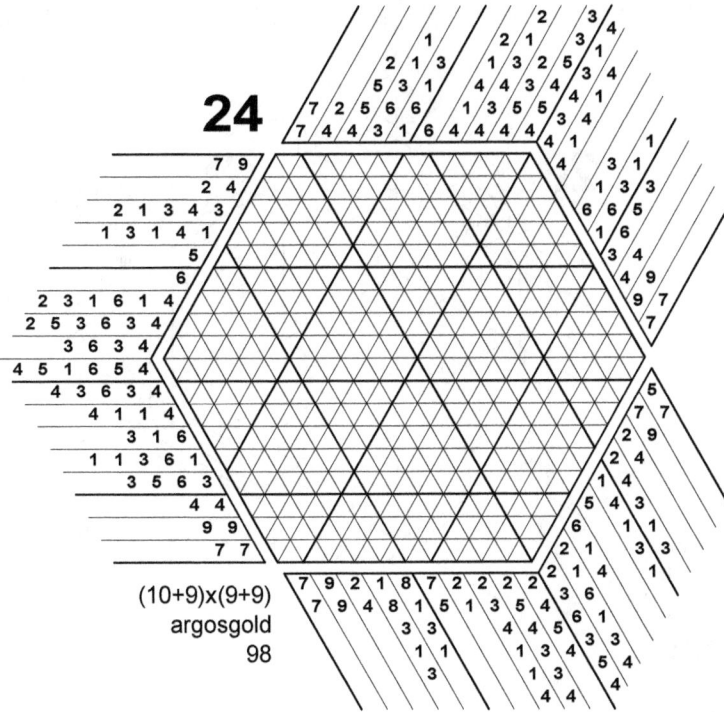

(10+9)x(9+9)
argosgold
98

25

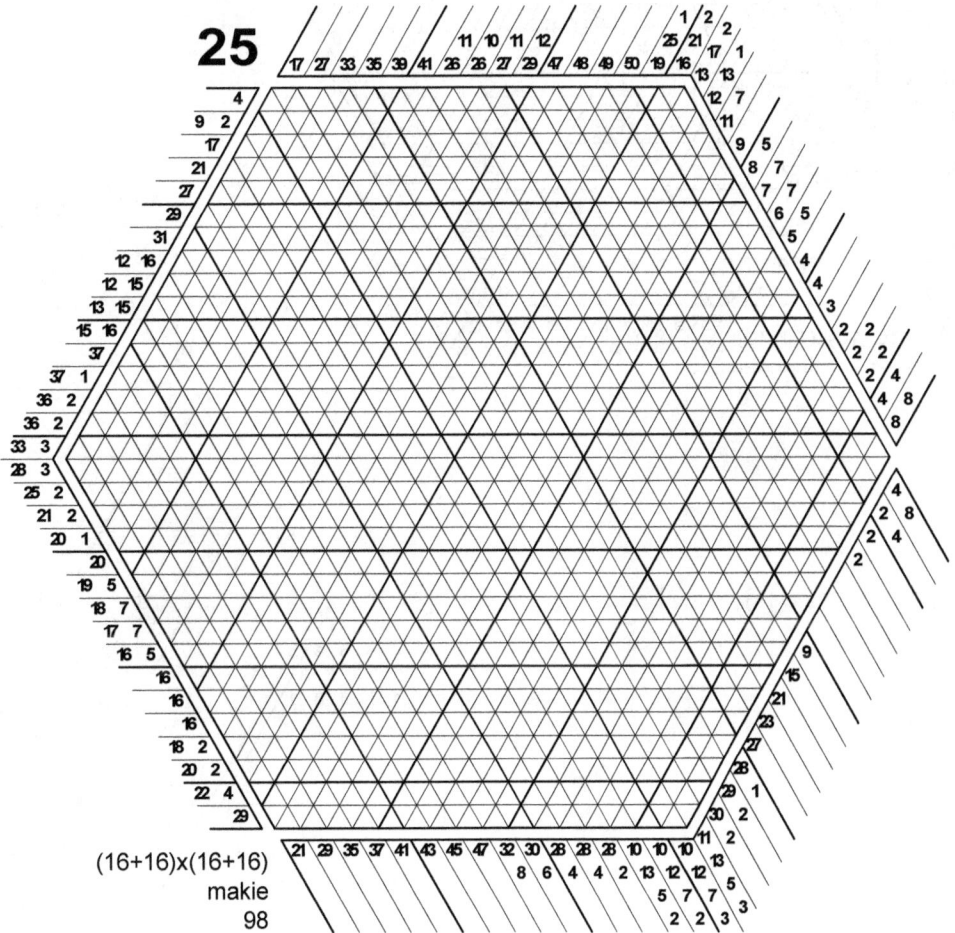

(16+16)x(16+16)
makie
98

26

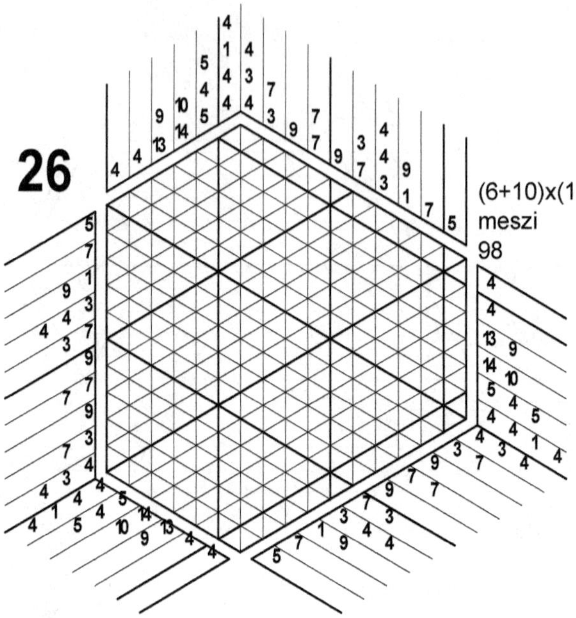

(6+10)x(10+6)
meszi
98

27

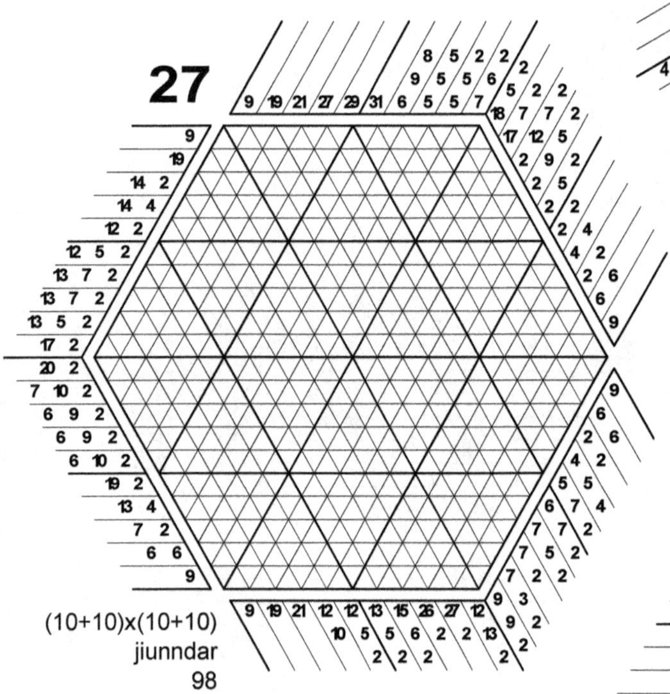

(10+10)x(10+10)
jiunndar
98

28

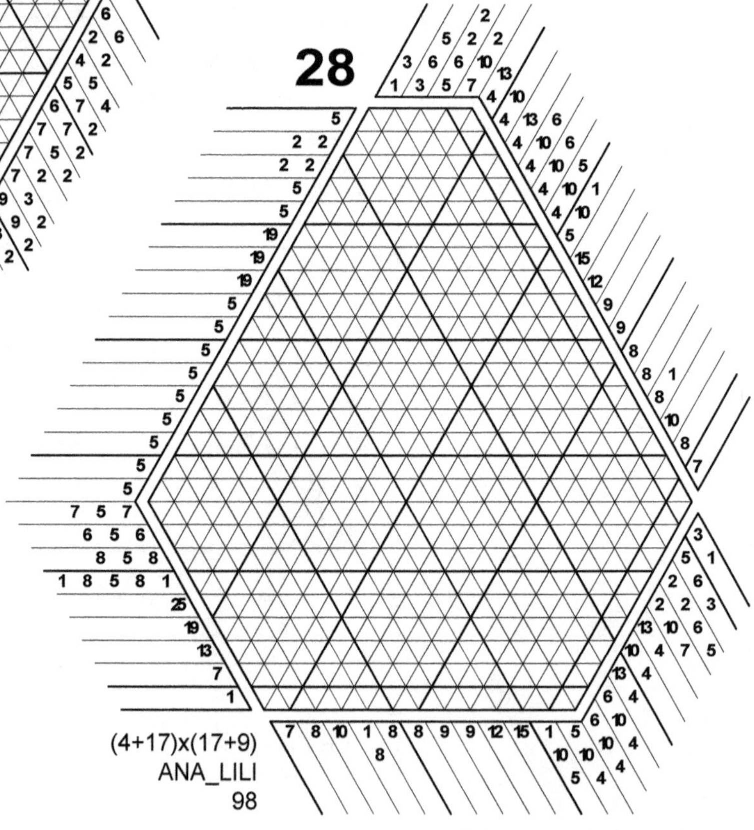

(4+17)x(17+9)
ANA_LILI
98

29

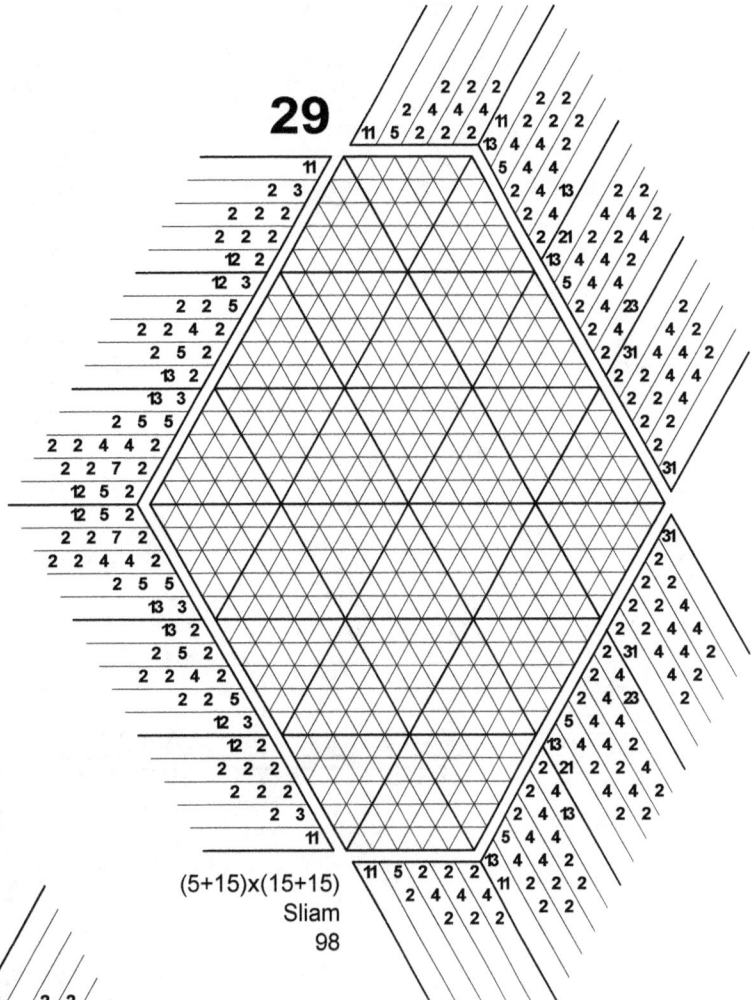

(5+15)x(15+15)
Sliam
98

30

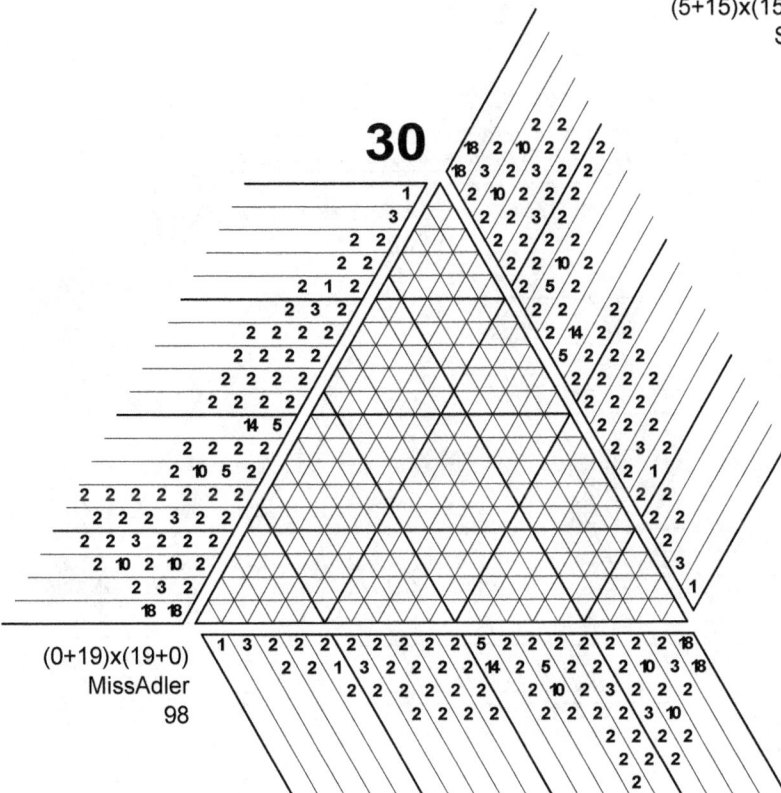

(0+19)x(19+0)
MissAdler
98

31

(15+18)x(19+9)
kayton
98

32

(17+12)x(12+10)
cekpompiere
98

33

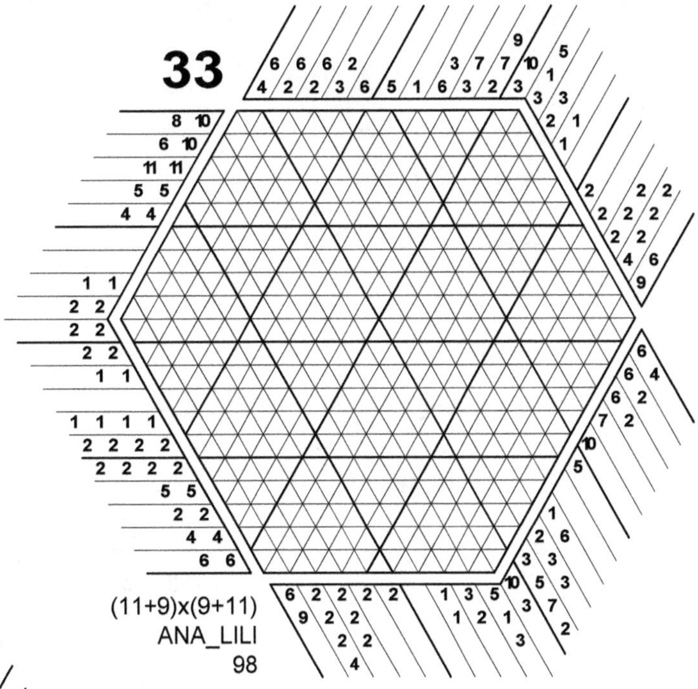

(11+9)x(9+11)
ANA_LILI
98

34

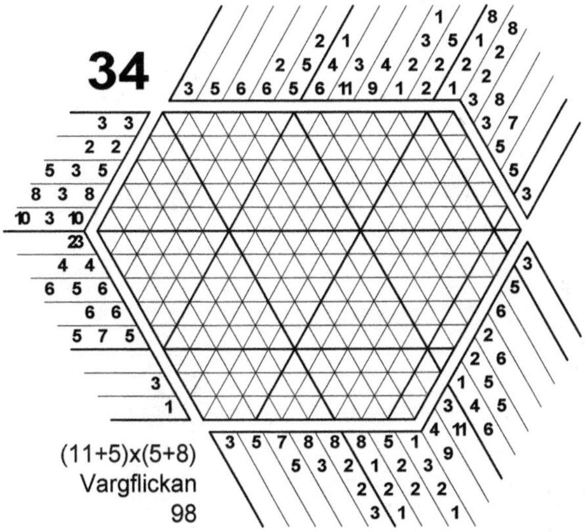

(11+5)x(5+8)
Vargflickan
98

35

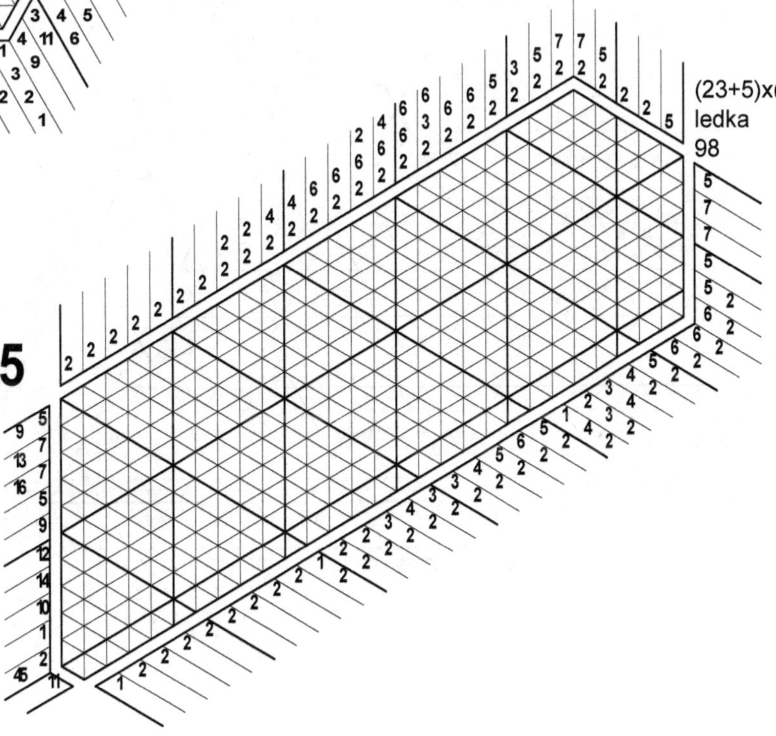

(23+5)x(10+1)
ledka
98

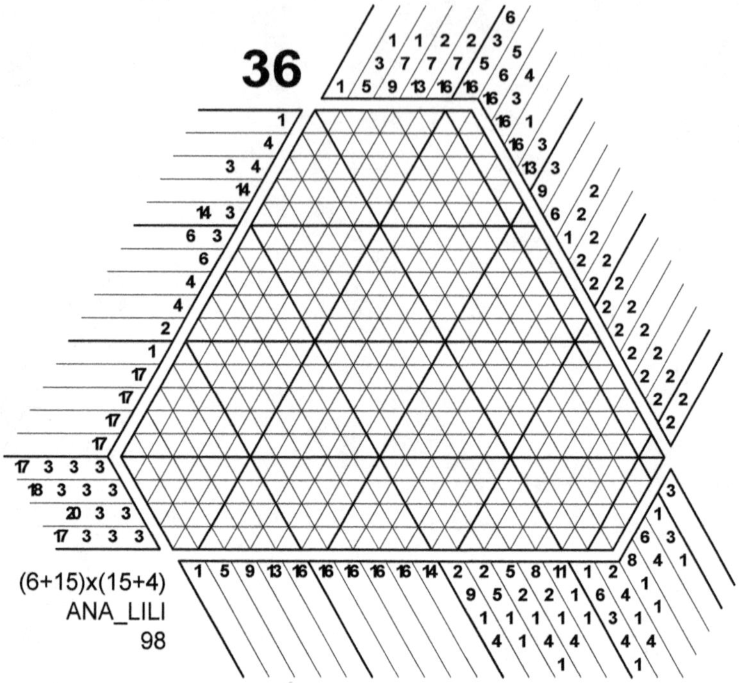

(6+15)x(15+4)
ANA_LILI
98

37

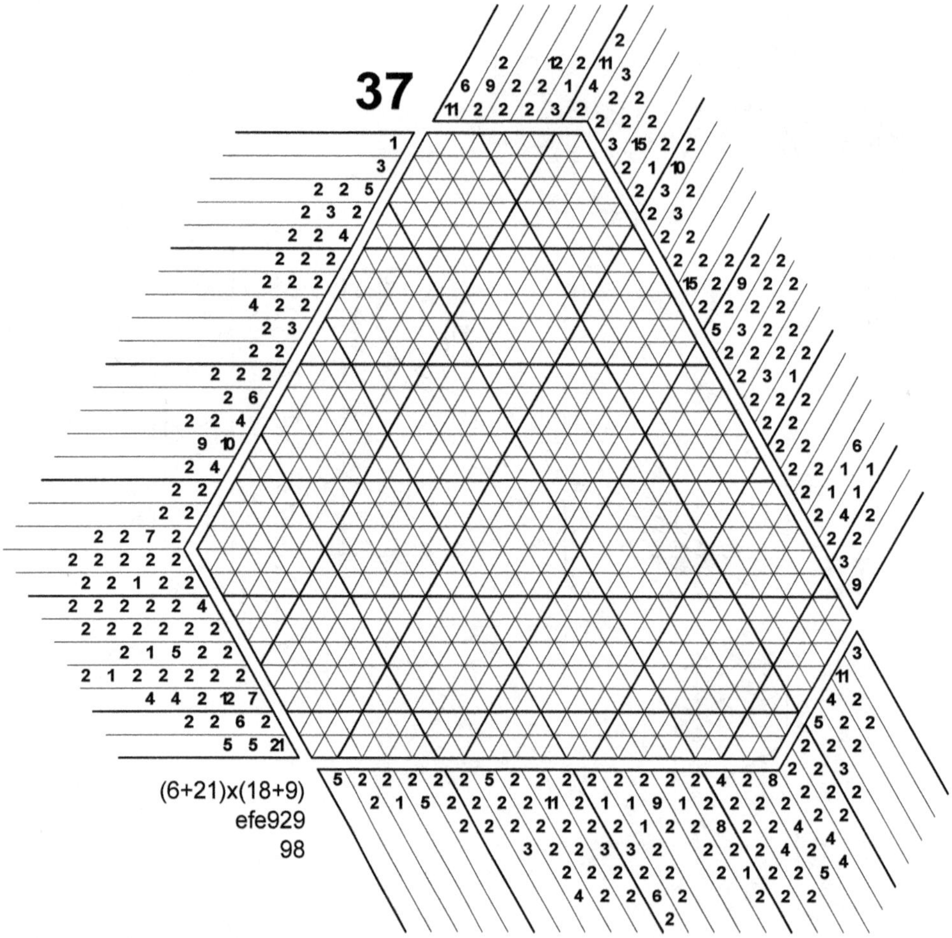

(6+21)x(18+9)
efe929
98

38

(31+13)x(19+25)
hertseltsur
99

39

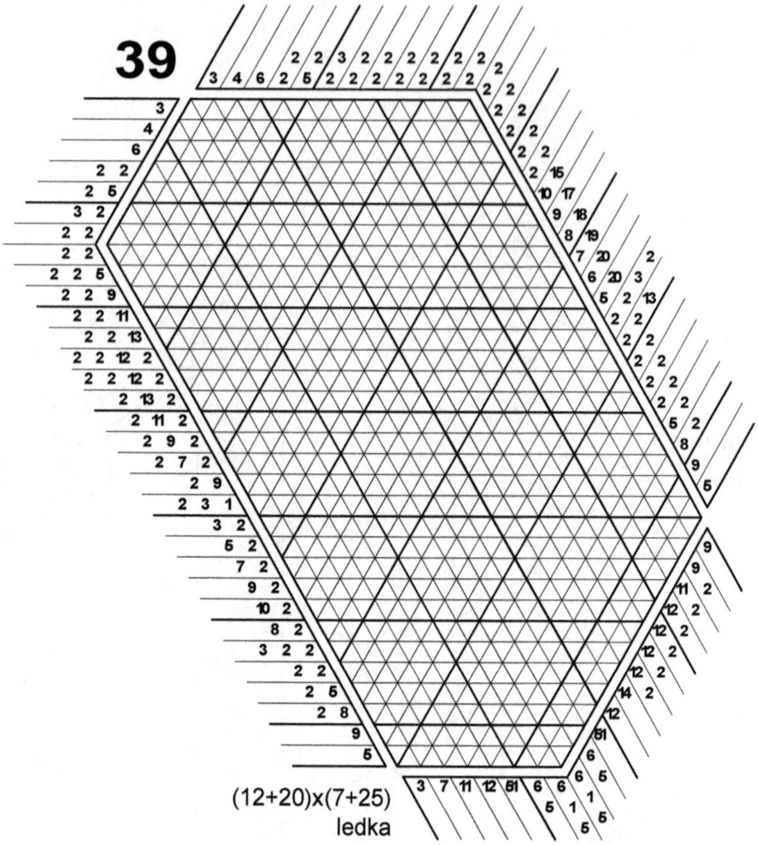

(12+20)x(7+25)
ledka
99

40

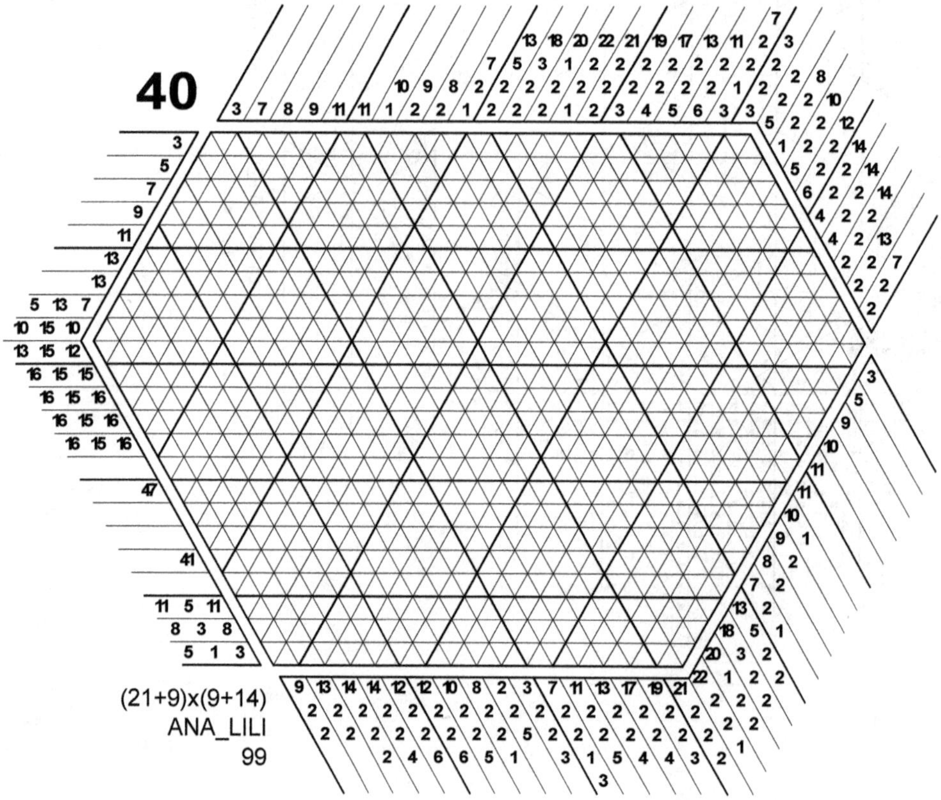

(21+9)x(9+14)
ANA_LILI
99

41

(7+7)x(12+11)
ANA_LILI
99

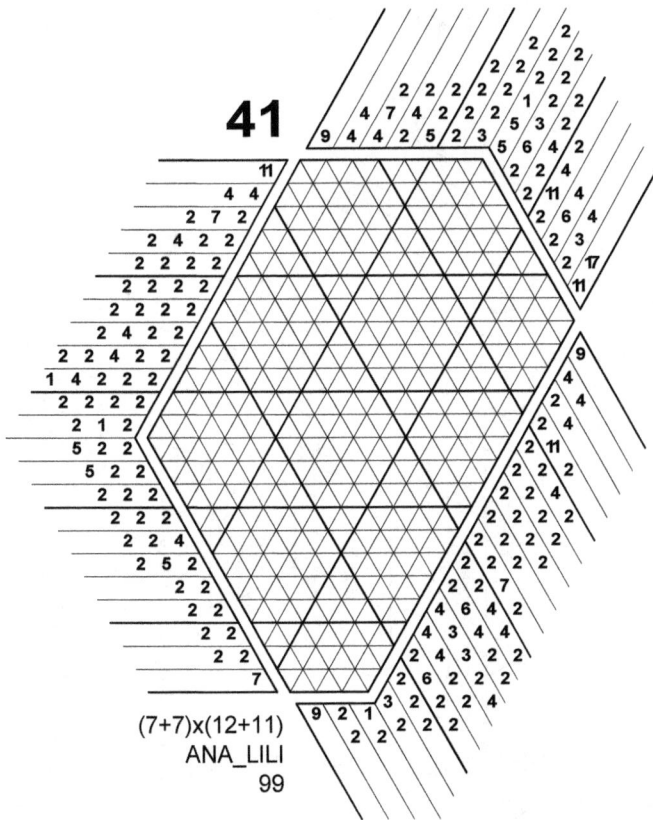

42

(11+12)x(14+11)
bedirhan
99

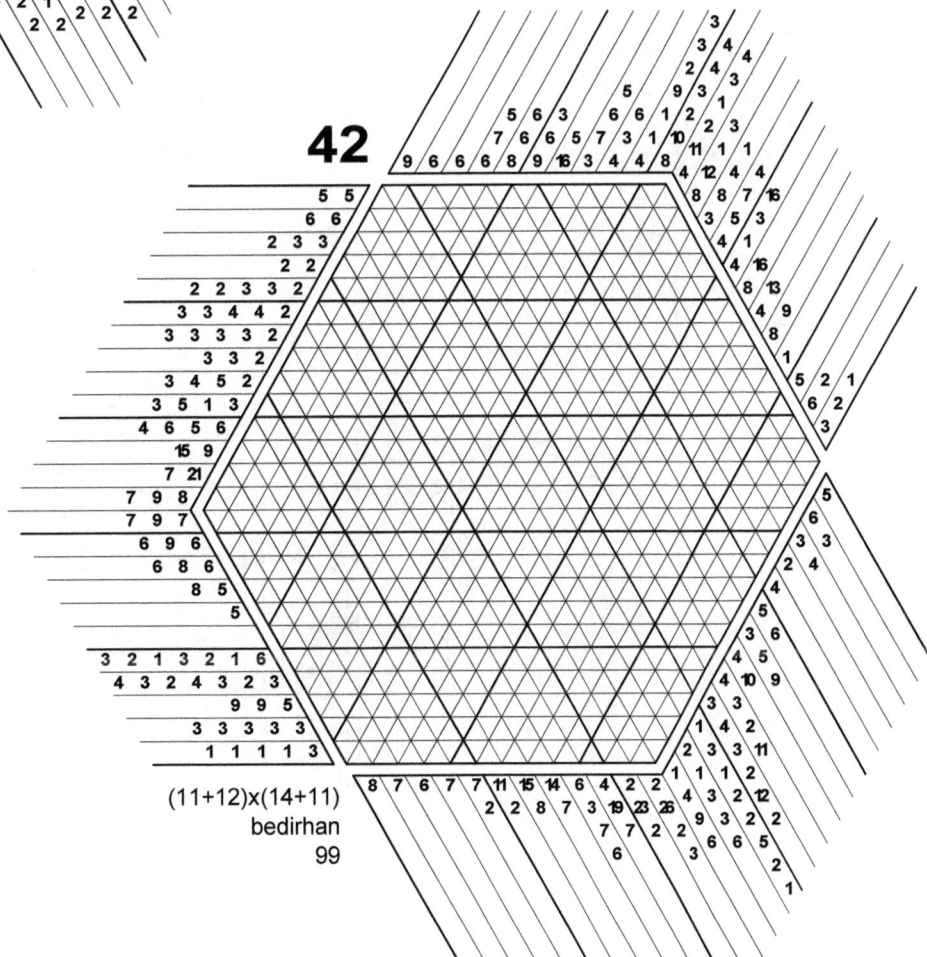

43

(22+22)x(23+21)
ANA_LILI
99

44

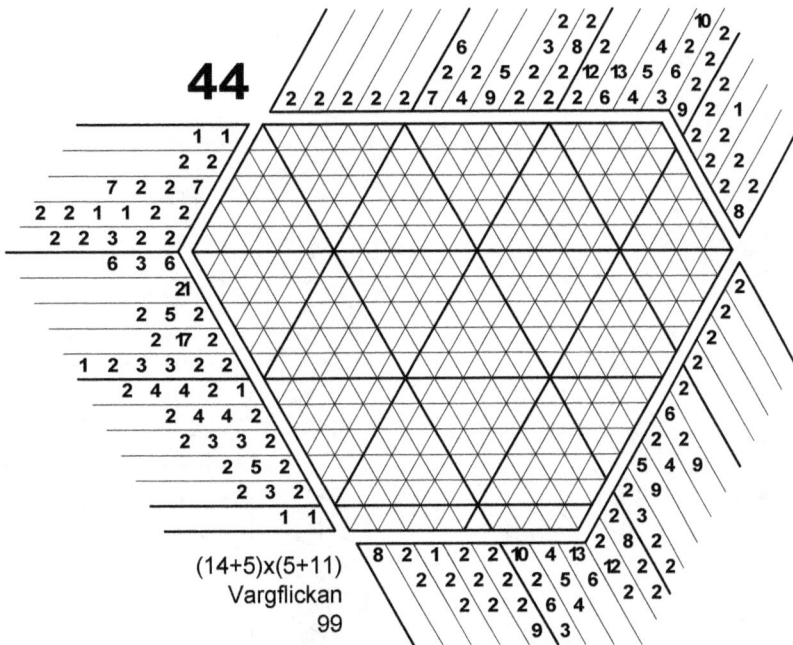

(14+5)x(5+11)
Vargflickan
99

45

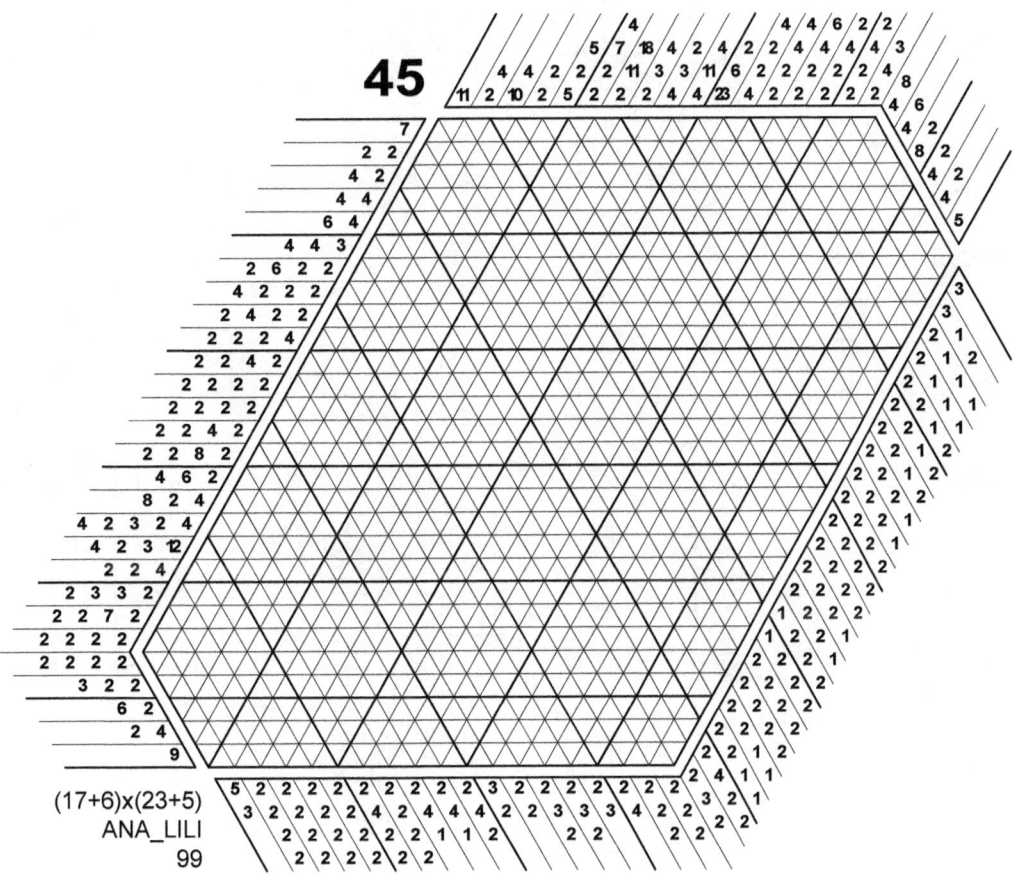

(17+6)x(23+5)
ANA_LILI
99

46

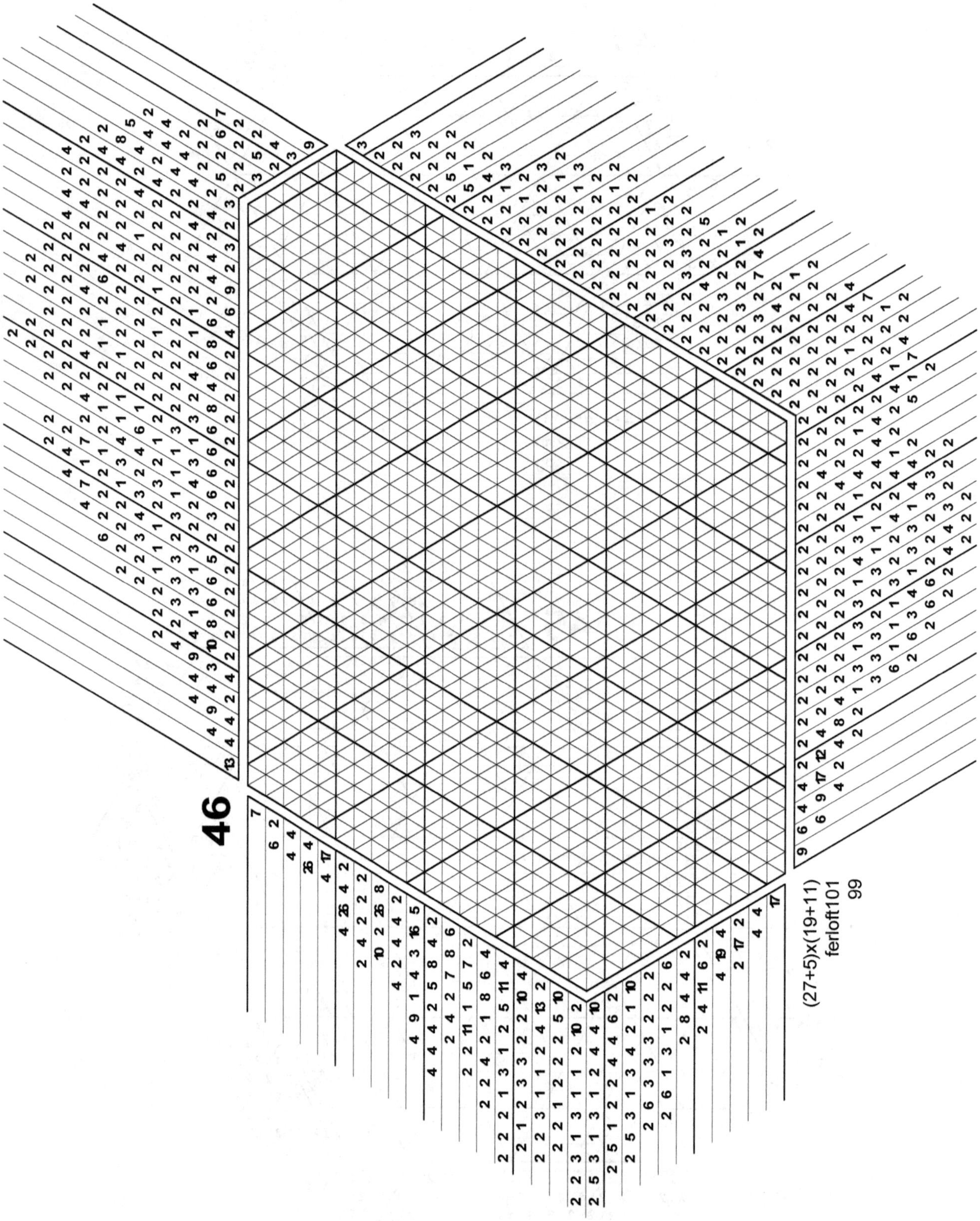

47

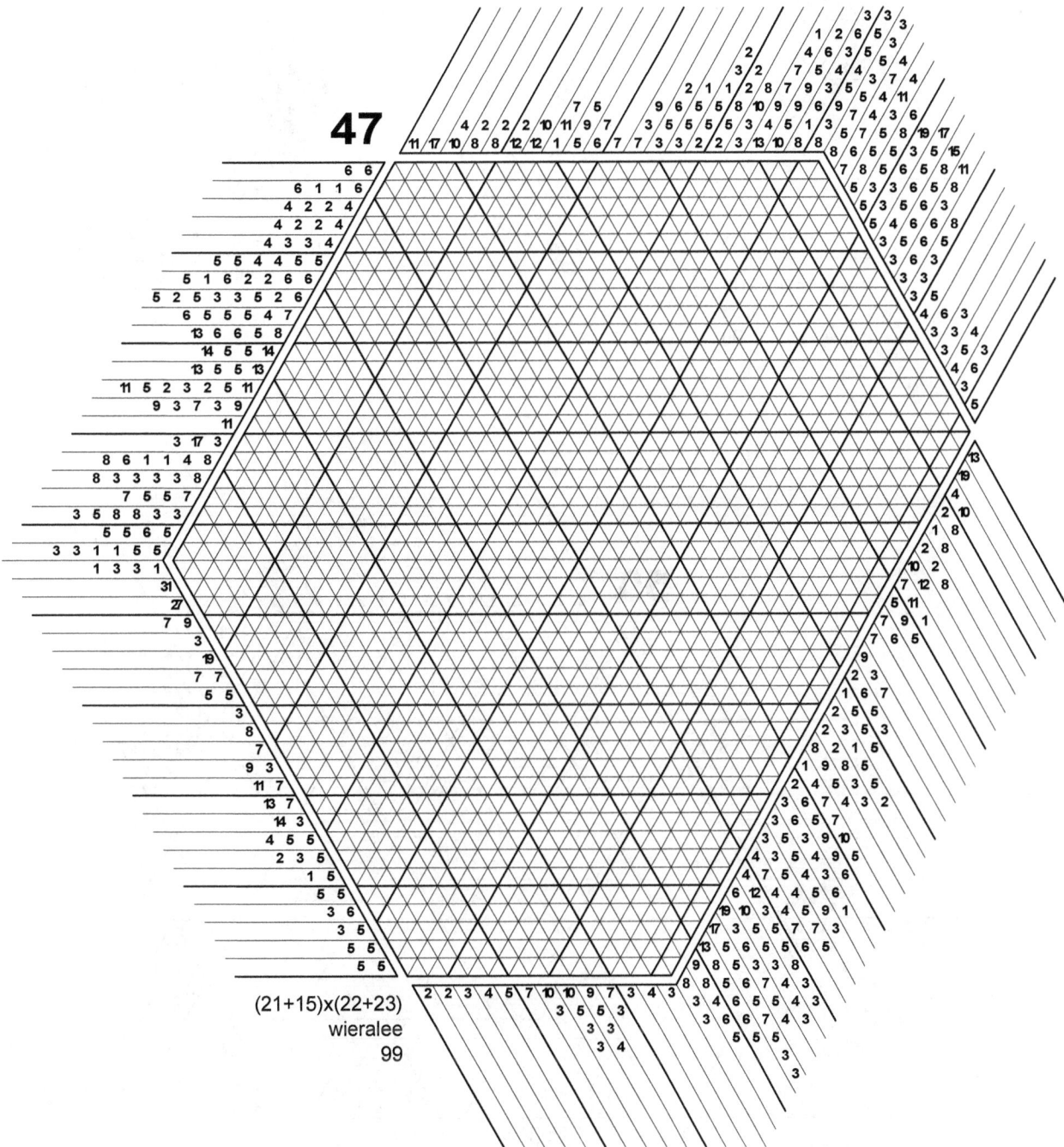

(21+15)x(22+23)
wieralee
99

48

(14+4)x(2+12)
myreg
99

49

(5+5)x(5+5)
zjmonty
99

50

(15+12)x(21+4)
bilbobilbo
99

51

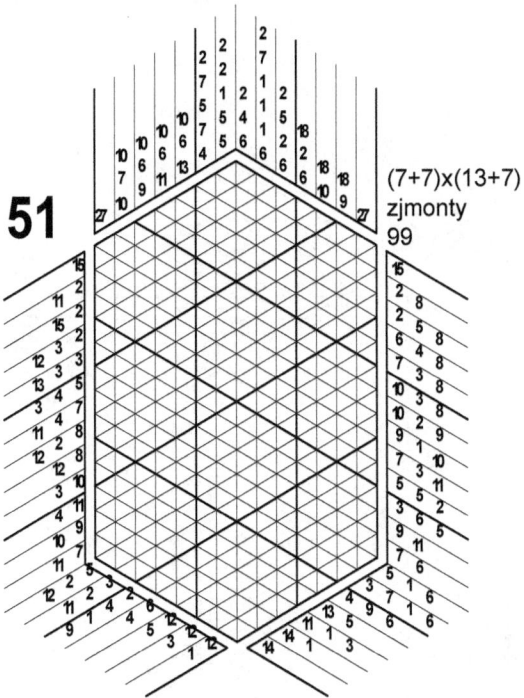

(7+7)x(13+7)
zjmonty
99

52

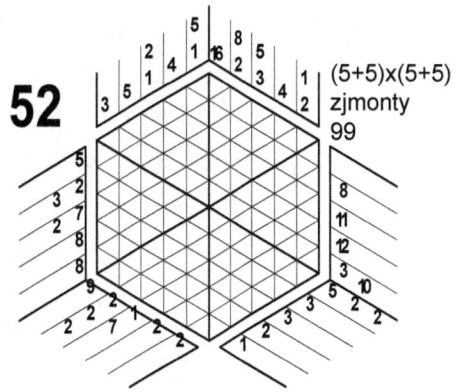

(5+5)x(5+5)
zjmonty
99

53

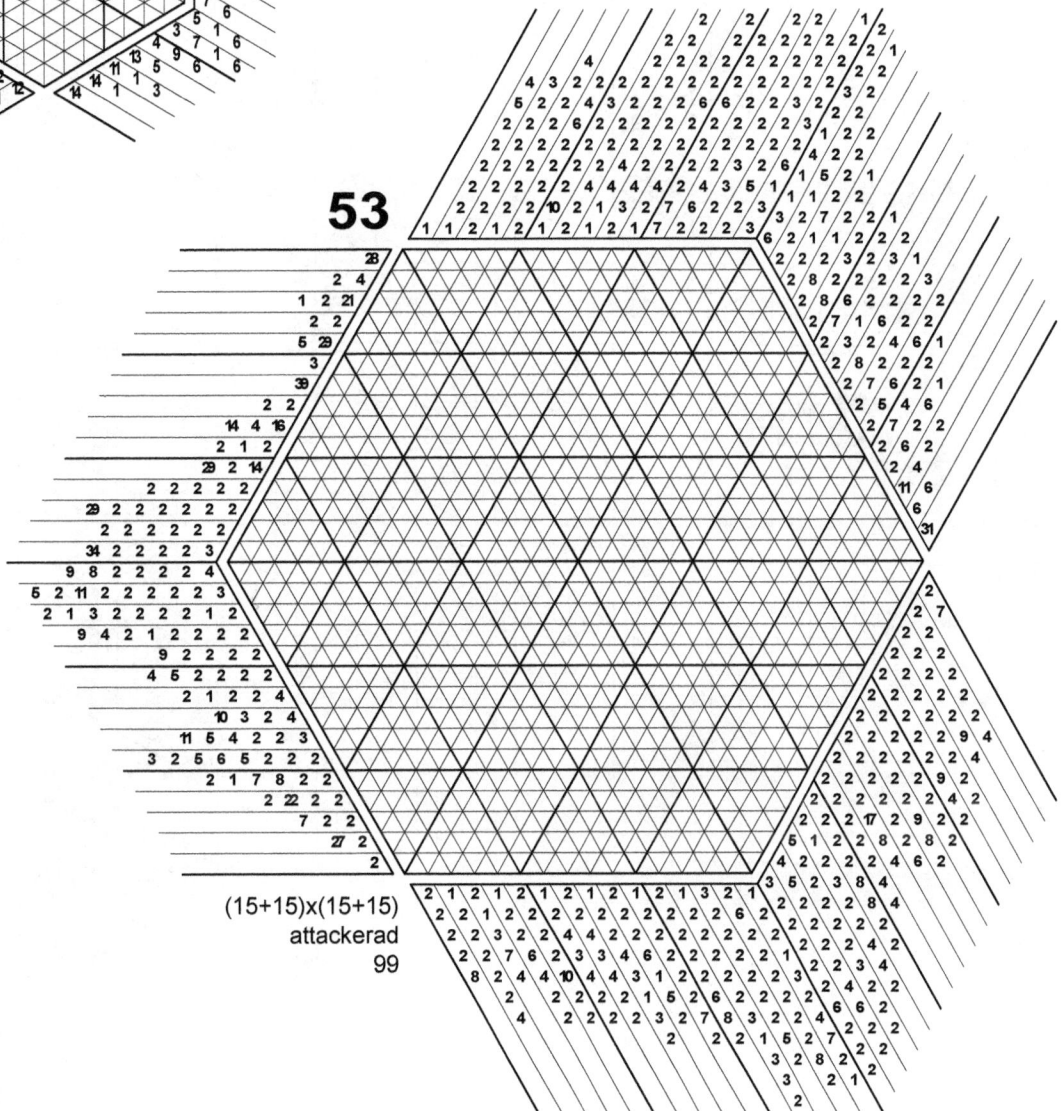

(15+15)x(15+15)
attackerad
99

54

(18+25)x(32+8)
ANA_LILI
99

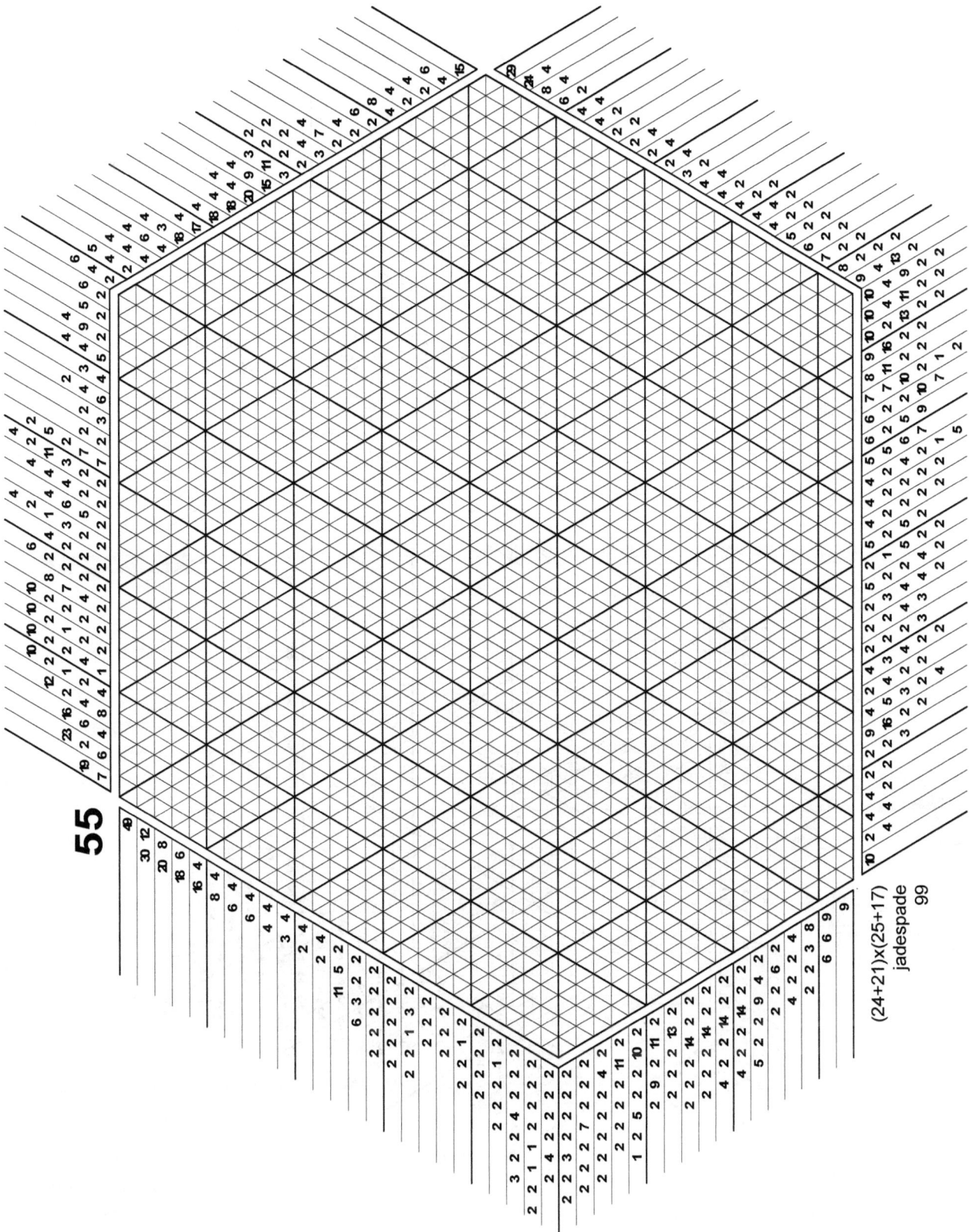

55

(24+21)x(25+17)
jadespade
99

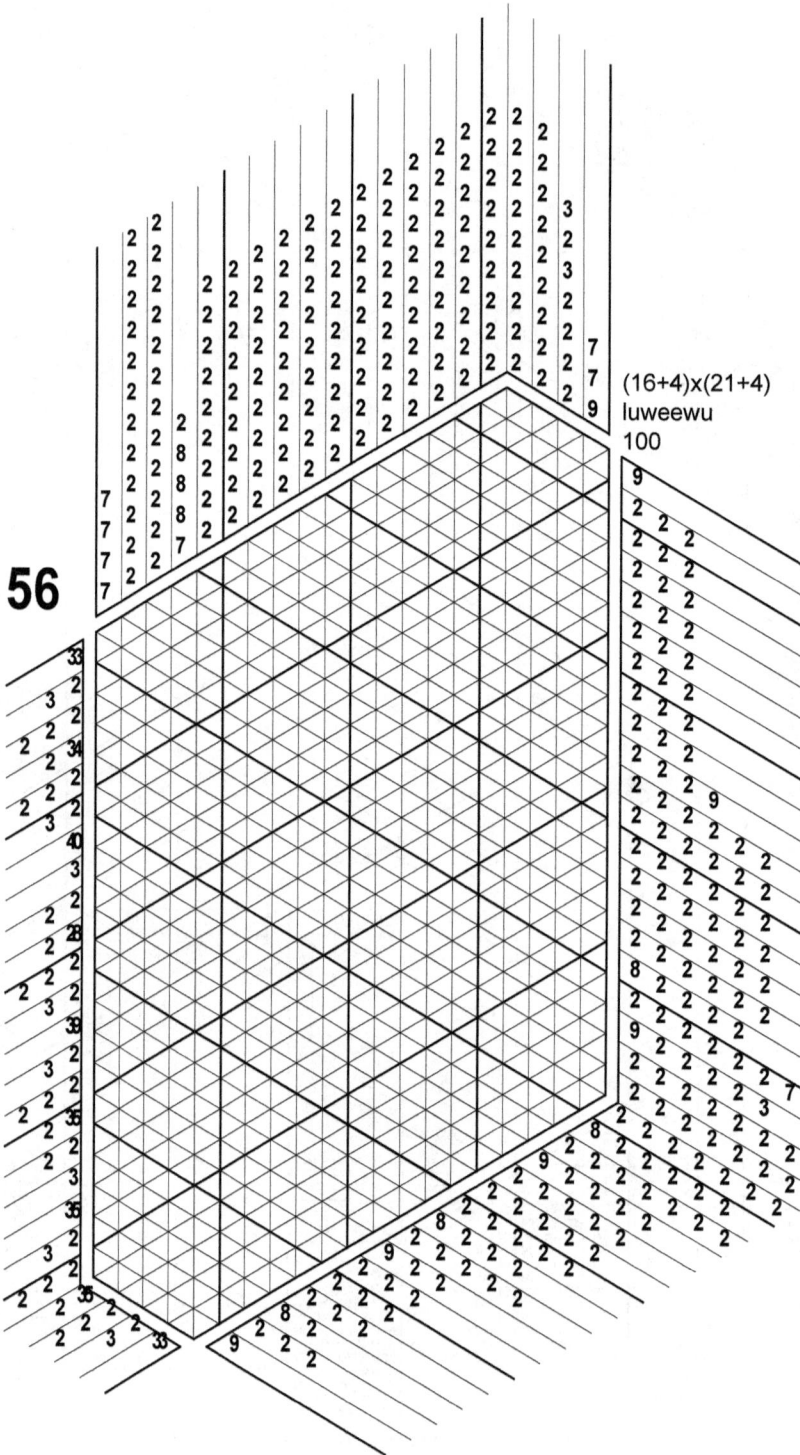

56

(16+4)x(21+4)
luweewu
100

57

(5+5)x(5+5)
hi19hi19
100

58

(15+15)x(15+15)
Elena_13
100

59

(13+7)x(7+5)
Vargflickan
100

60

(22+14)x(27+8)
lenchik
100

61

(12+27)x(25+11)
ANA_LILI
100

(32+13)x(13+29)
hertseltsur
100

62

63

(20+15)x(30+10)
amsterdam
100

64

(5+11)x(17+6)
ANA_LILI
100

65

(8+17)x(13+14)
Aranlyde
100

67

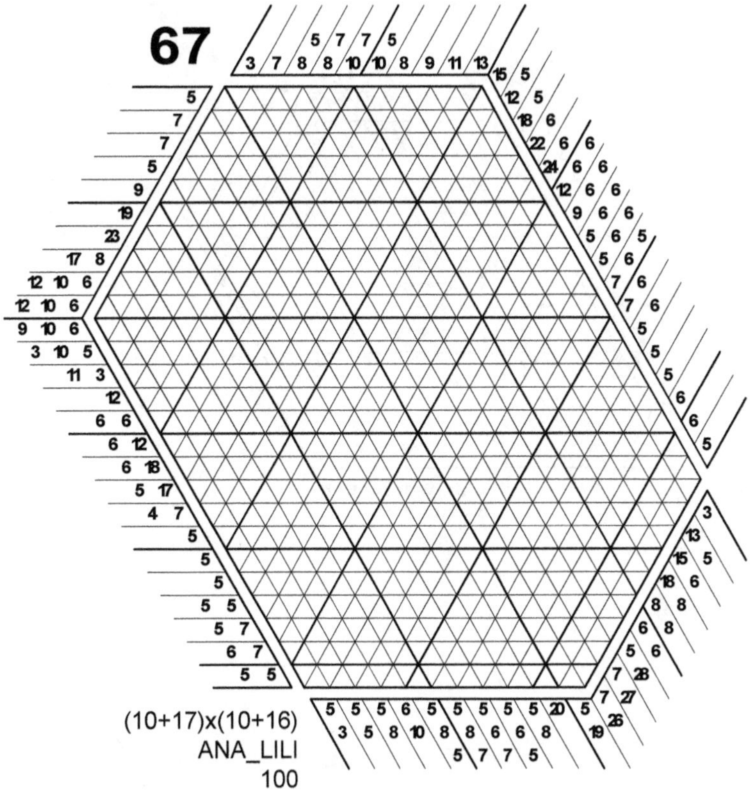

(10+17)x(10+16)
ANA_LILI
100

66

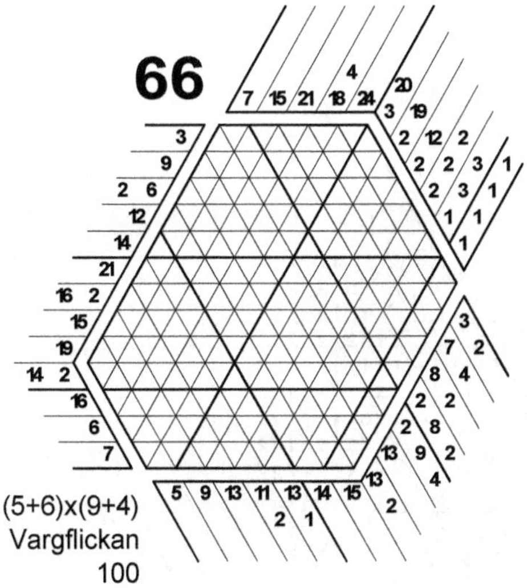

(5+6)x(9+4)
Vargflickan
100

68

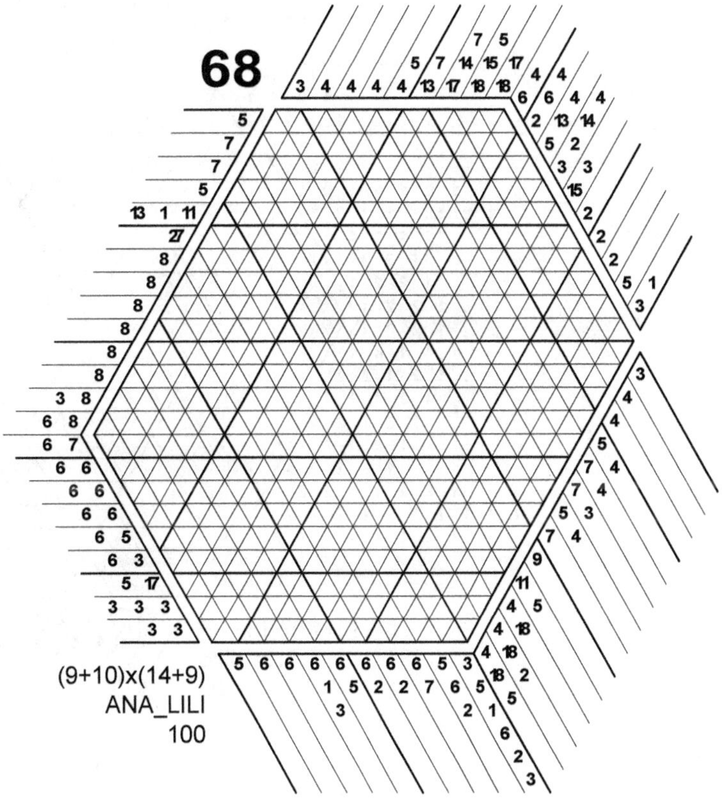

(9+10)x(14+9)
ANA_LILI
100

69

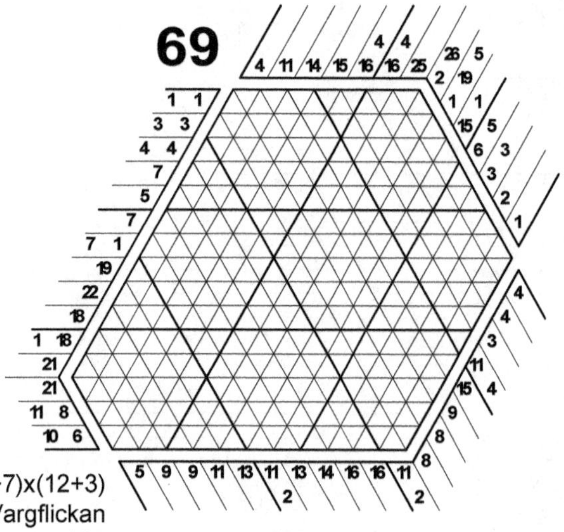

(7+7)x(12+3)
Vargflickan
100

70

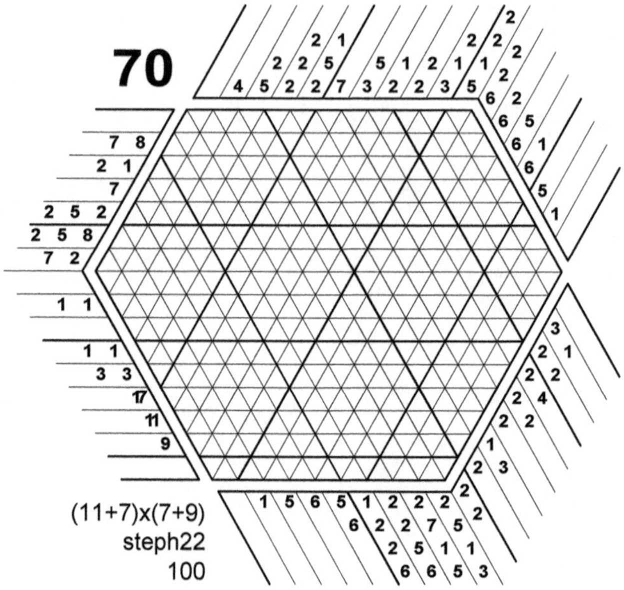

(11+7)x(7+9)
steph22
100

71

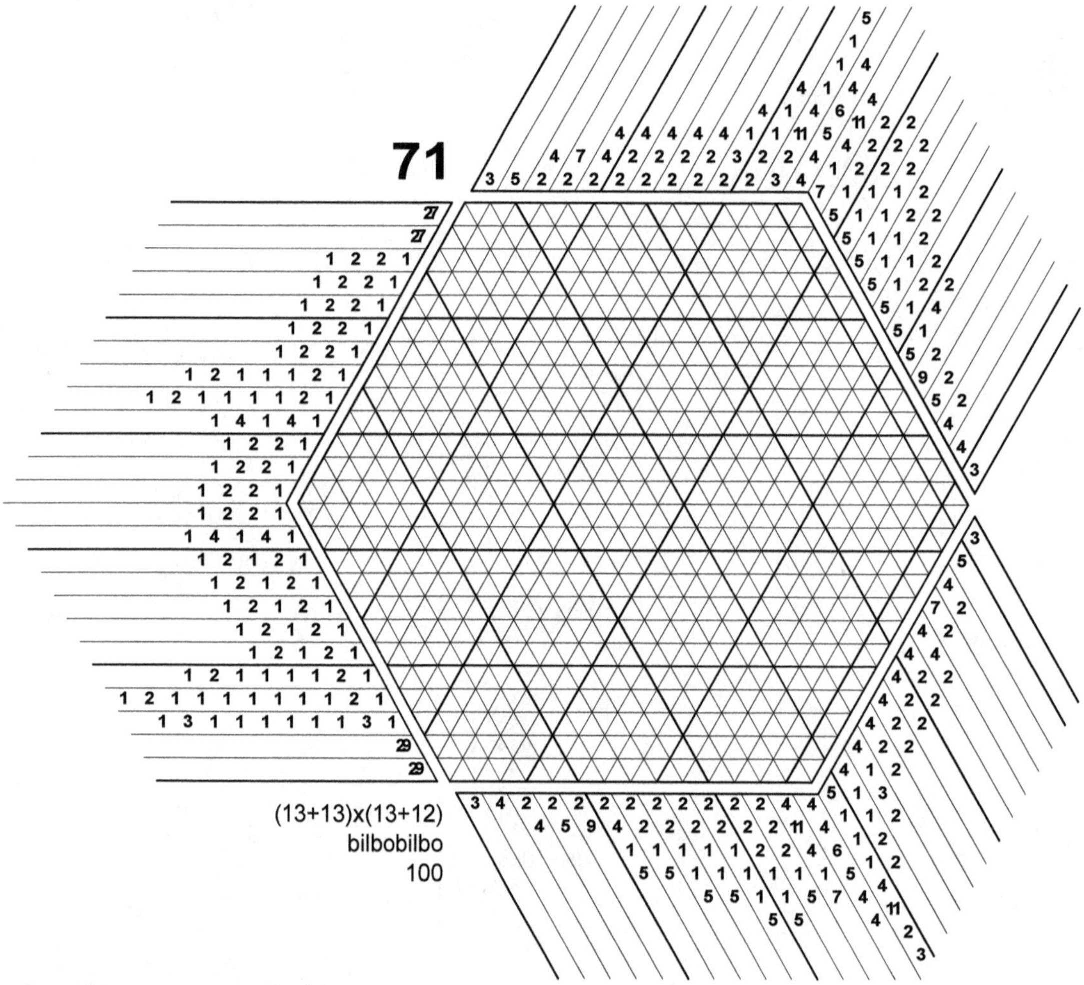

(13+13)x(13+12)
bilbobilbo
100

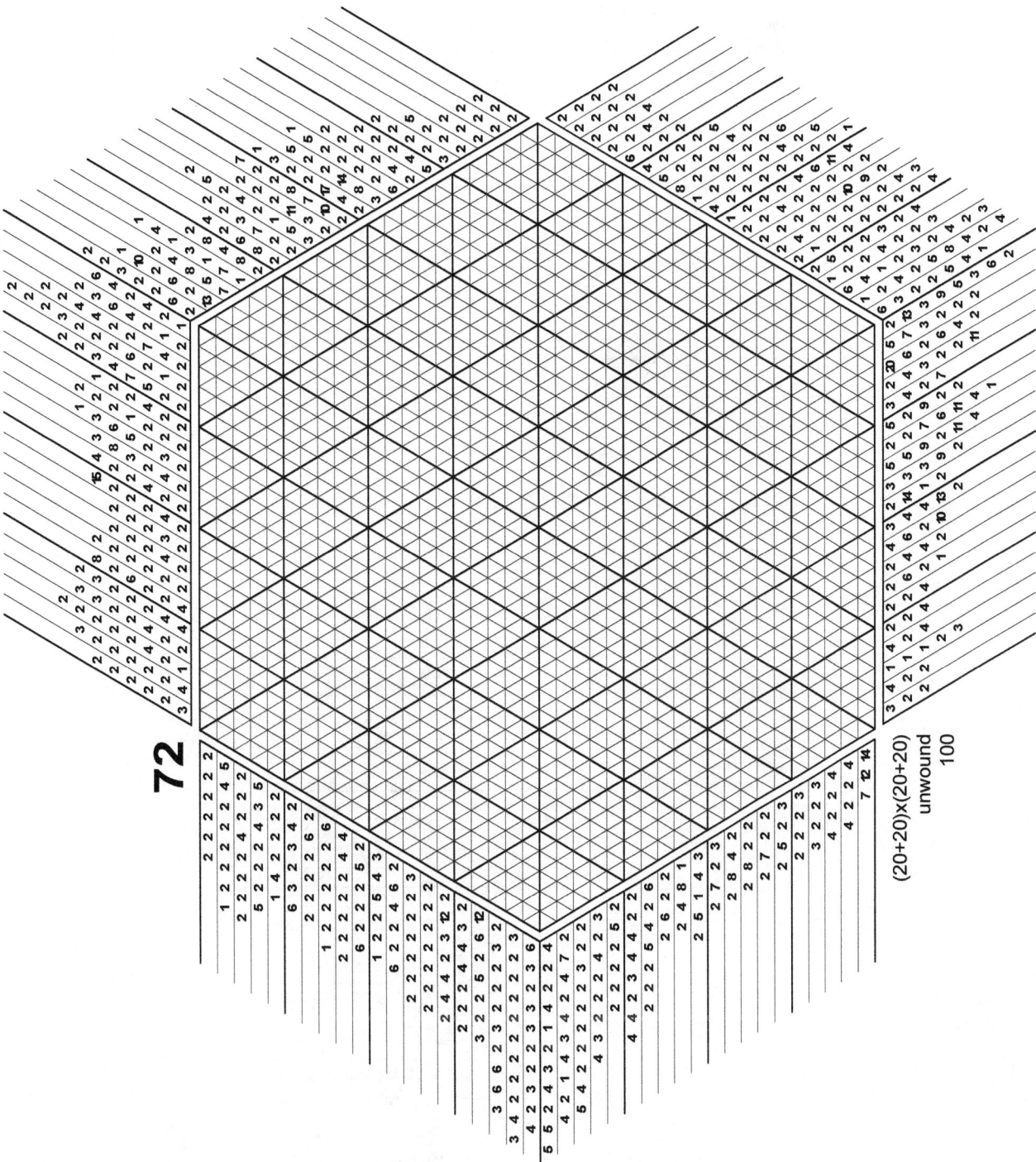

72

(20+20)x(20+20)
unwound
100

73

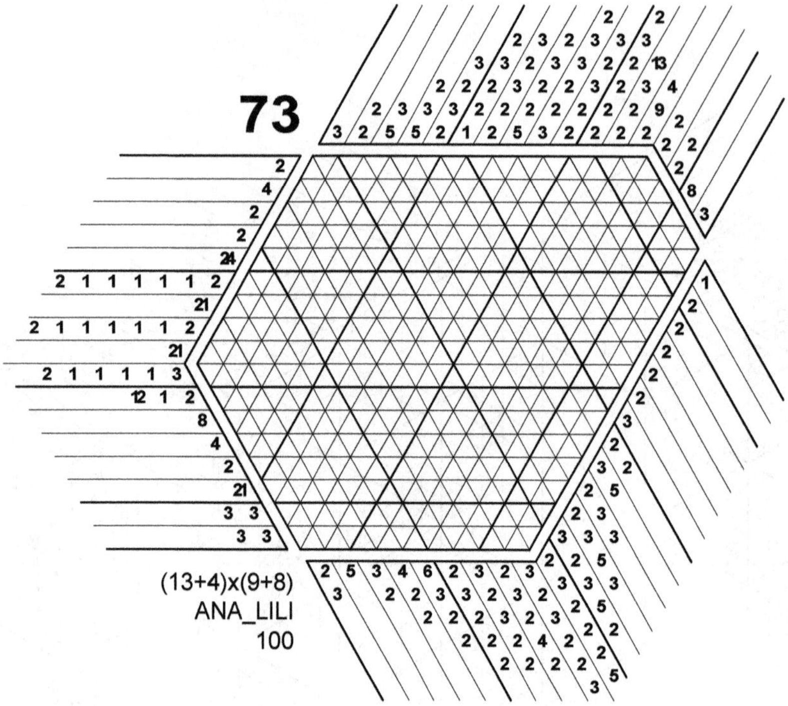

(13+4)x(9+8)
ANA_LILI
100

74

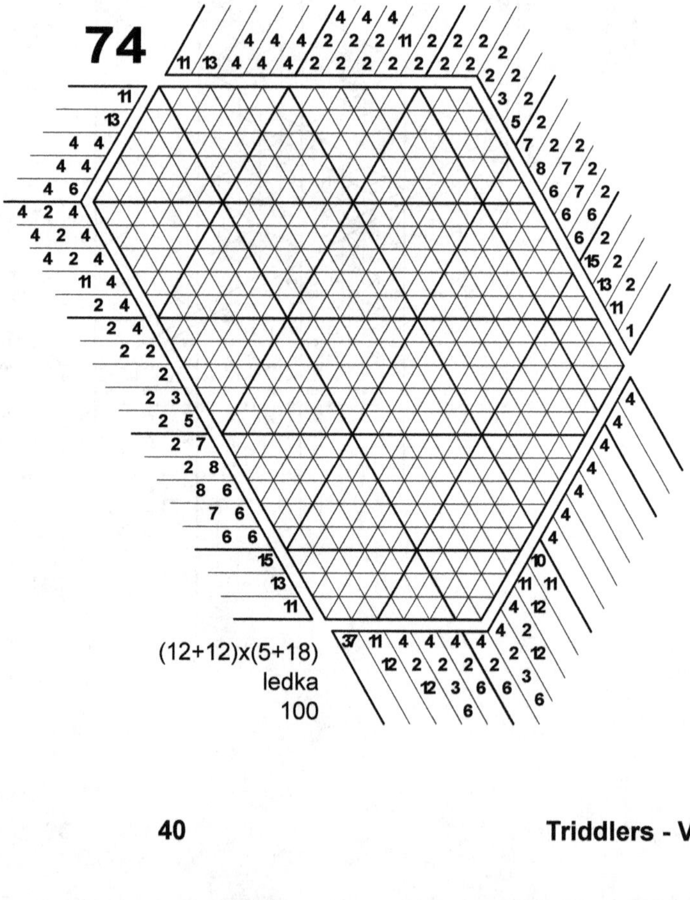

(12+12)x(5+18)
ledka
100

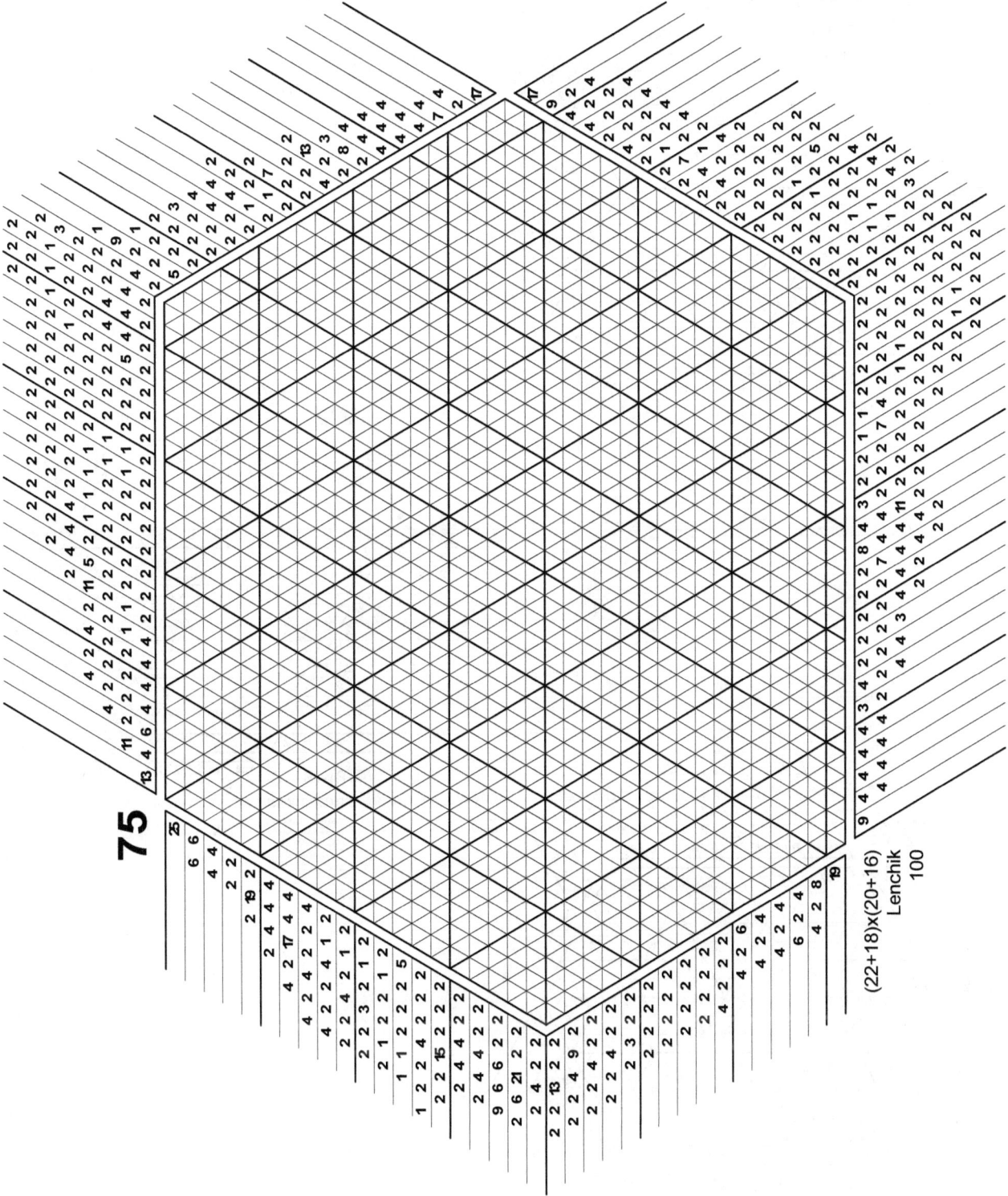

75

(22+18)x(20+16)
Lenchik
100

76

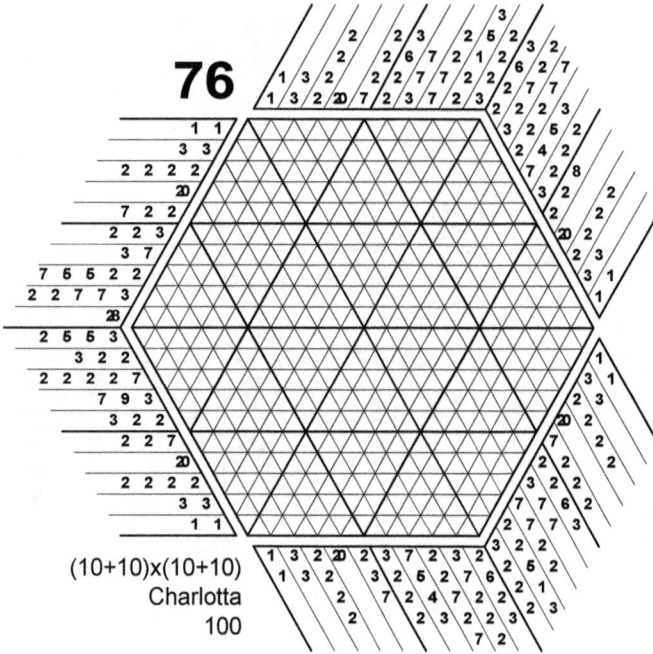

(10+10)x(10+10)
Charlotta
100

77

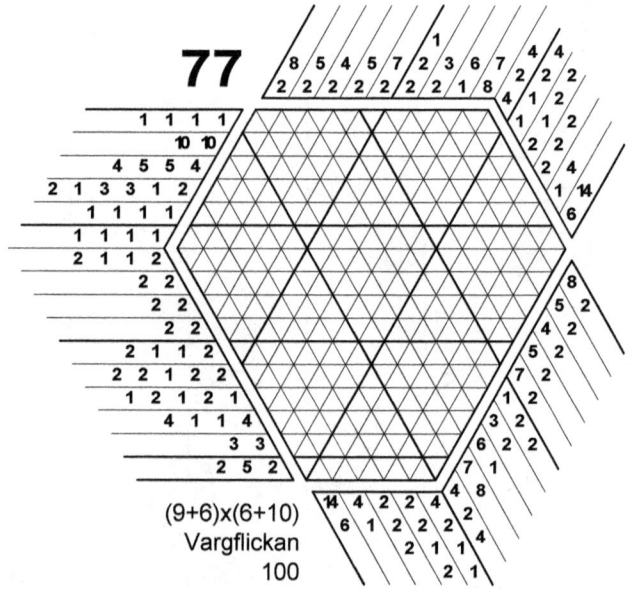

(9+6)x(6+10)
Vargflickan
100

78

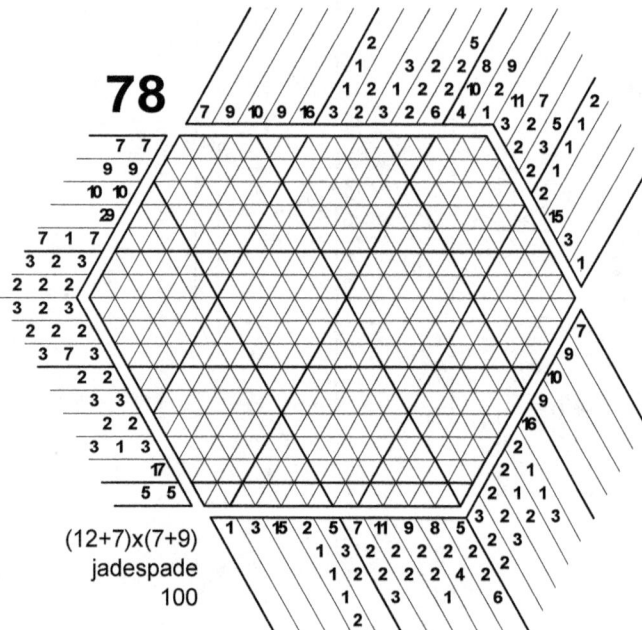

(12+7)x(7+9)
jadespade
100

79

(4+32)x(30+9)
ledka
100

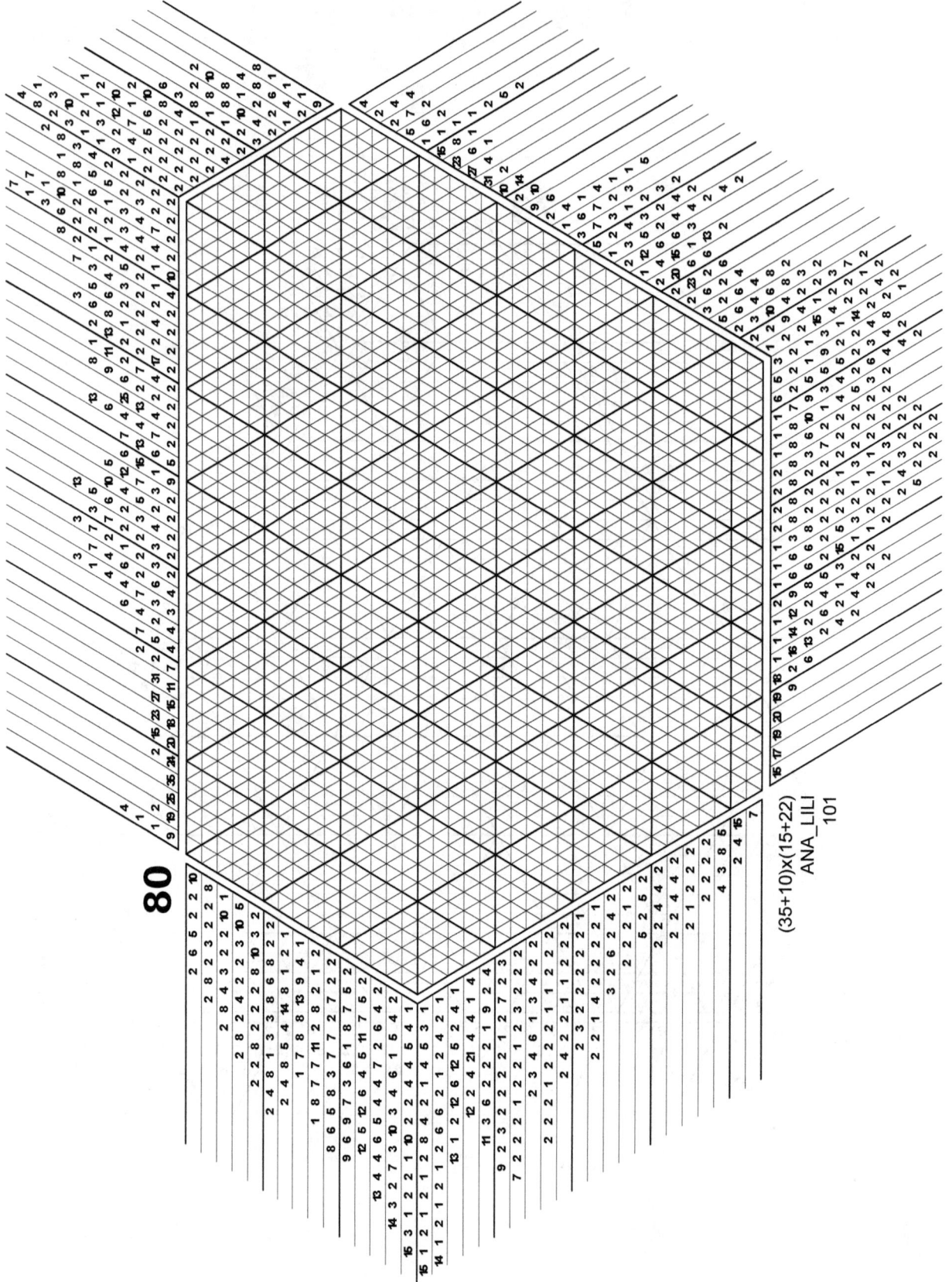

(35+10)x(15+22)
ANA_LILI
_101

80

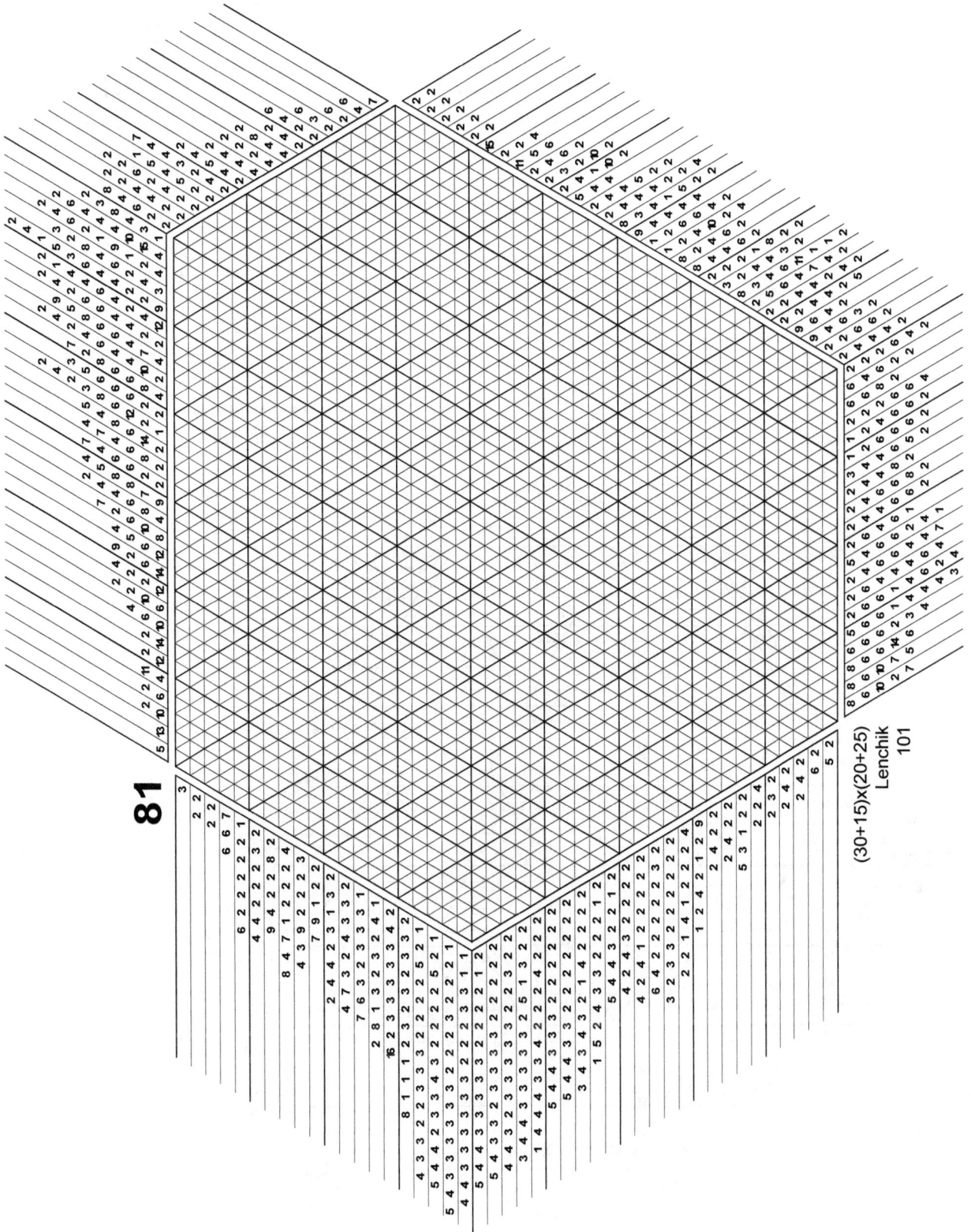

81

(30+15)x(20+25)
Lenchik
101

82

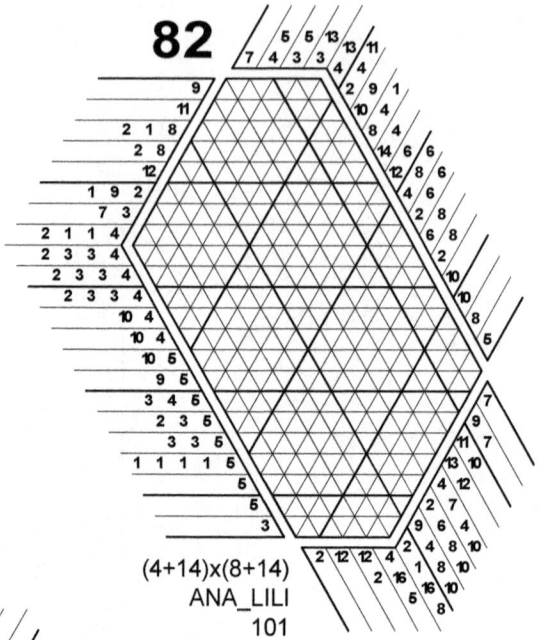

(4+14)x(8+14)
ANA_LILI
101

83

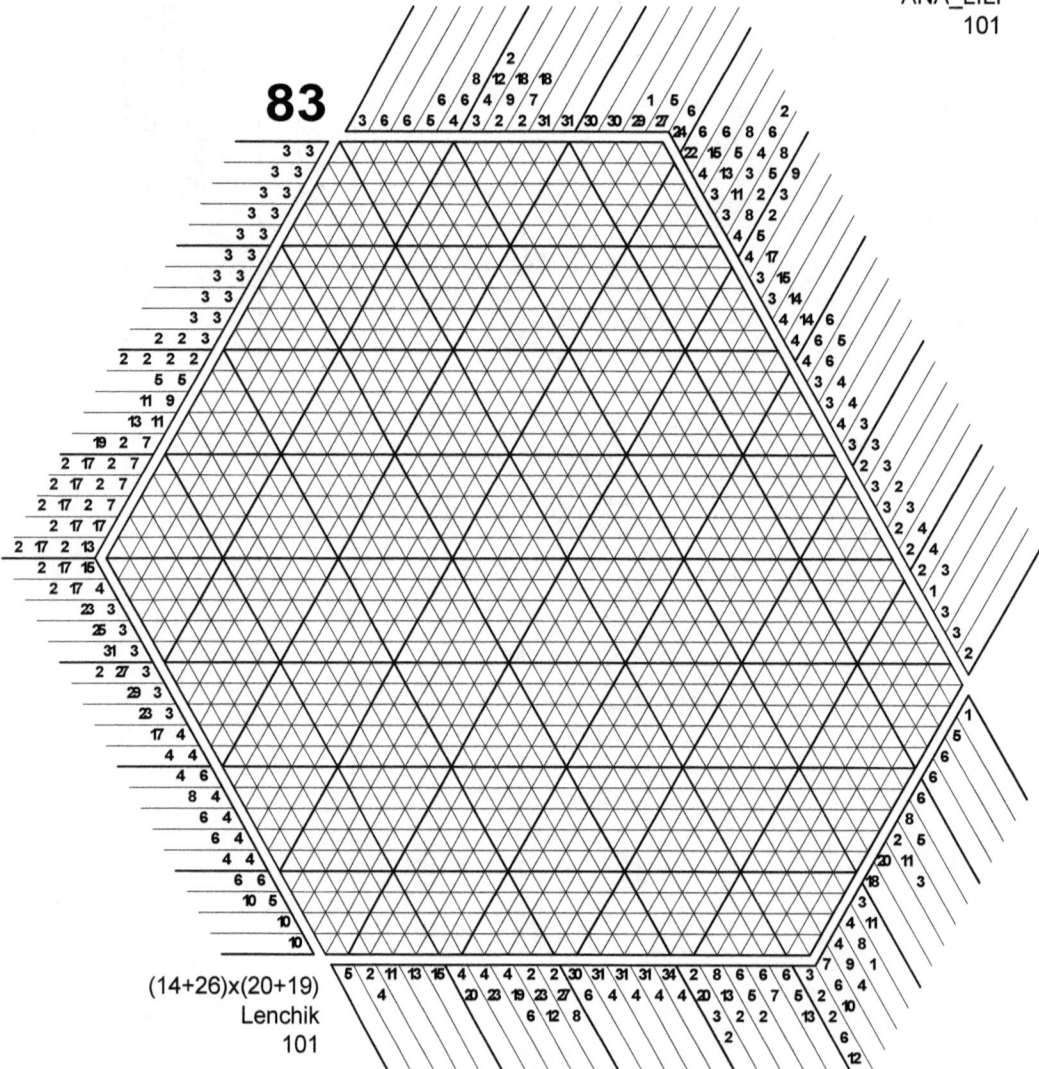

(14+26)x(20+19)
Lenchik
101

84

(9+7)x(7+5)
madziasek
101

85

(20+19)x(27+10)
Lenchik
101

86

(11+12)x(10+13)
ANA_LILI
101

87

(9+15)x(17+5)
hentron
101

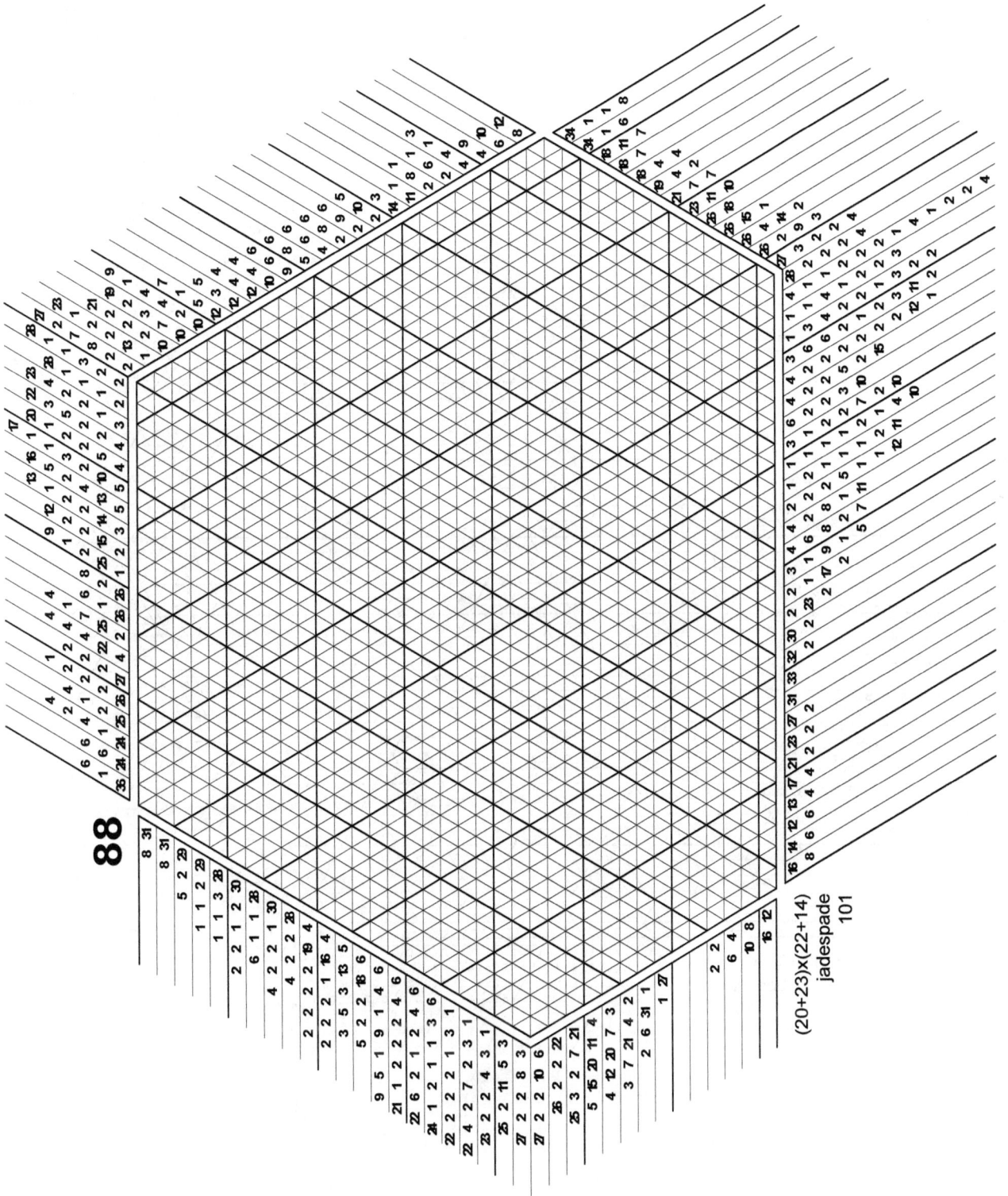

88

(20+23)x(22+14)
jadespade
101

89

(26+13)x(24+17)
Lenchik
101

90

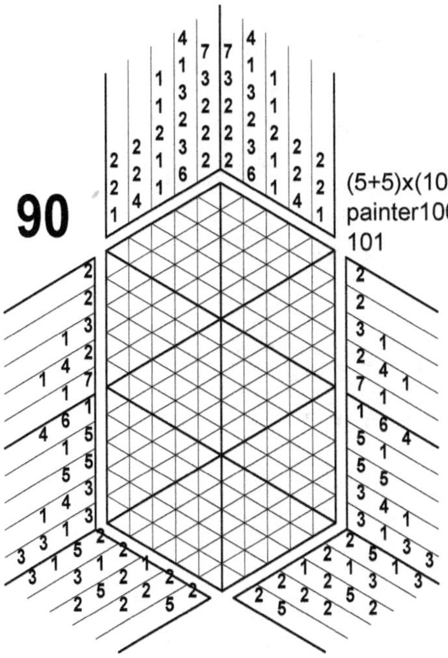

(5+5)x(10+5)
painter100
101

91

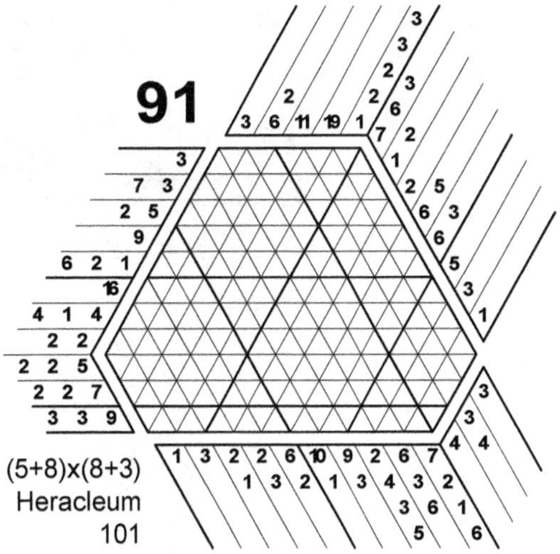

(5+8)x(8+3)
Heracleum
101

92

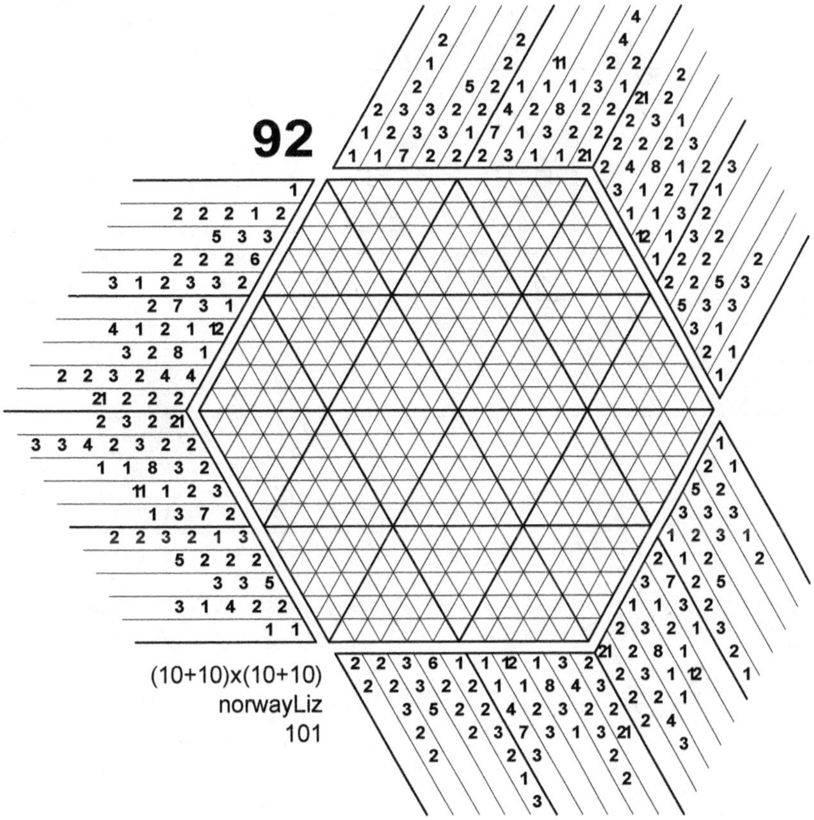

(10+10)x(10+10)
norwayLiz
101

93

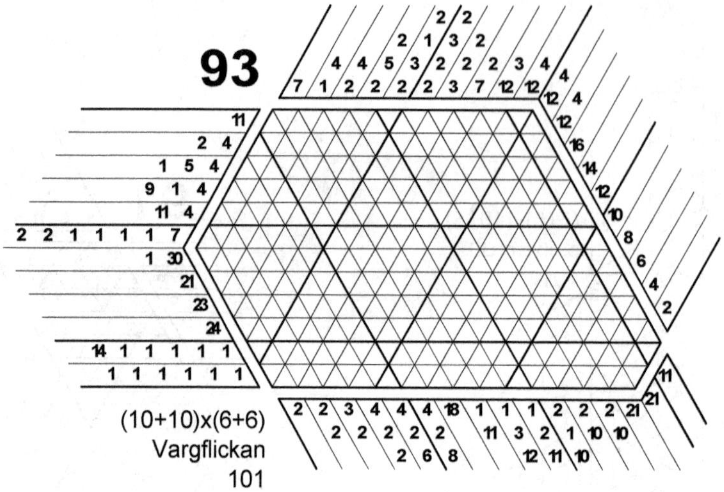

(10+10)x(6+6)
Vargflickan
101

94

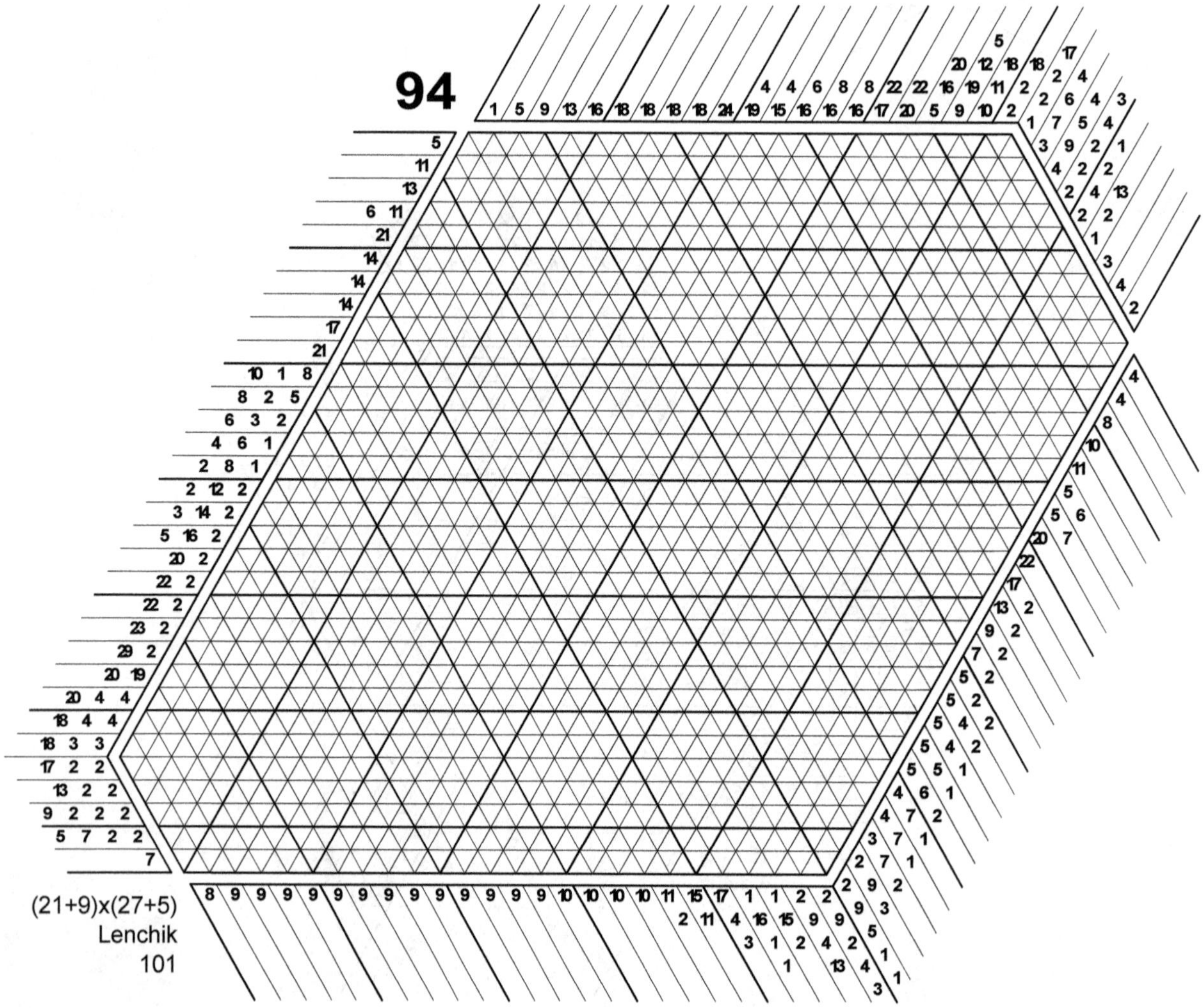

(21+9)x(27+5)
Lenchik
101

95

(16+17)x(17+17)
jadespade
101

96

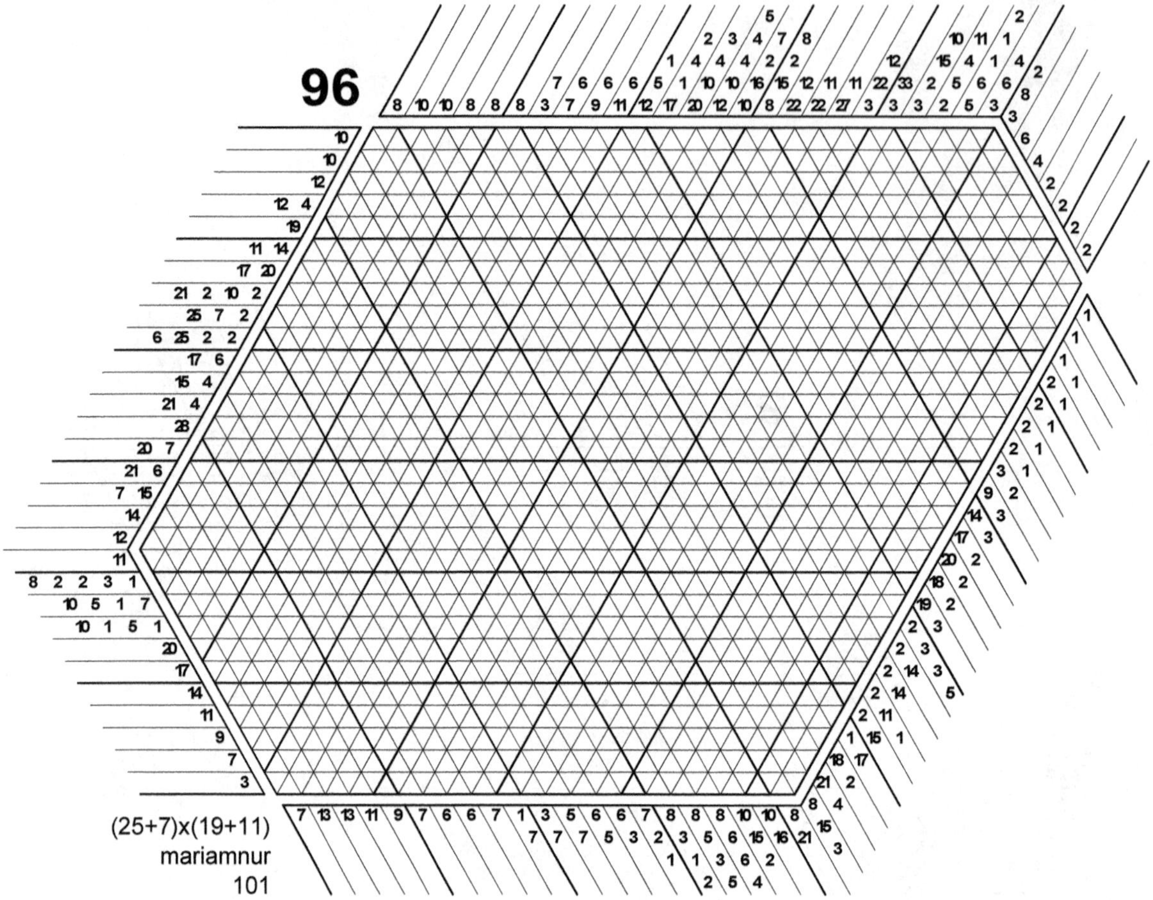

(25+7)x(19+11)
mariamnur
101

97

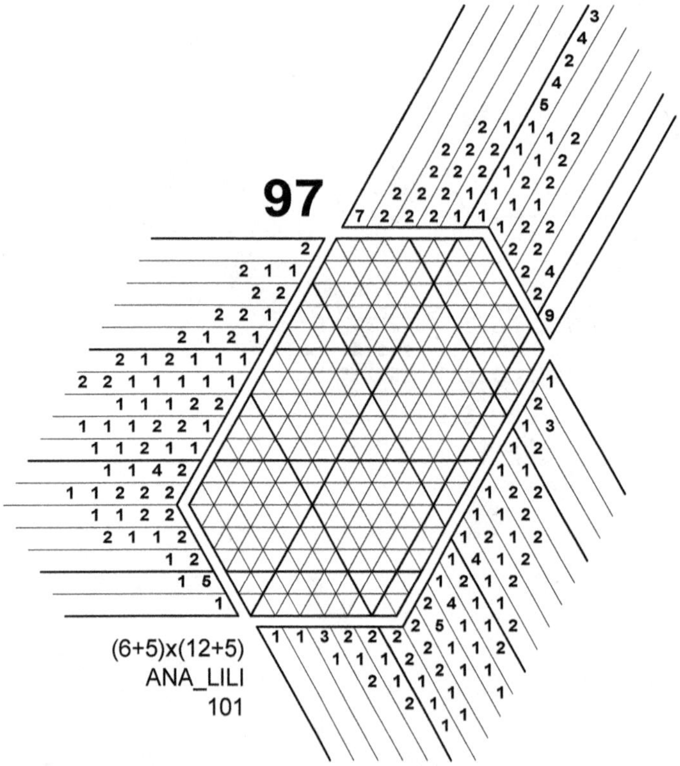

(6+5)x(12+5)
ANA_LILI
101

98

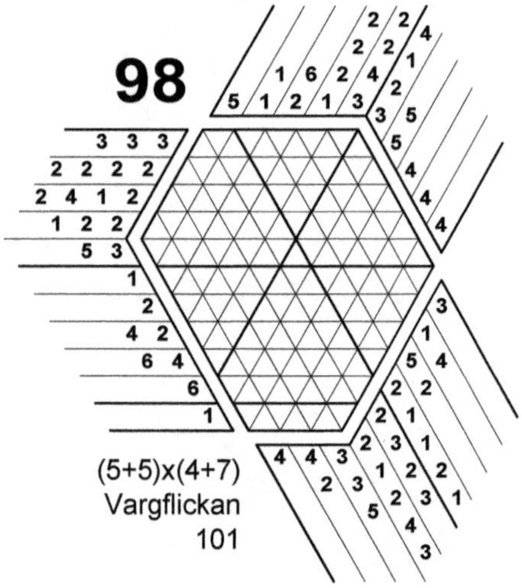

(5+5)x(4+7)
Vargflickan
101

99

(25+6)x(25+10)
ANA_LILI
_101

100

(10+3)x(22+2)
ledka
101

101

(10+10)x(6+14)
malymichu
101

102

(30+15)x(15+27)
hertseltsur
101

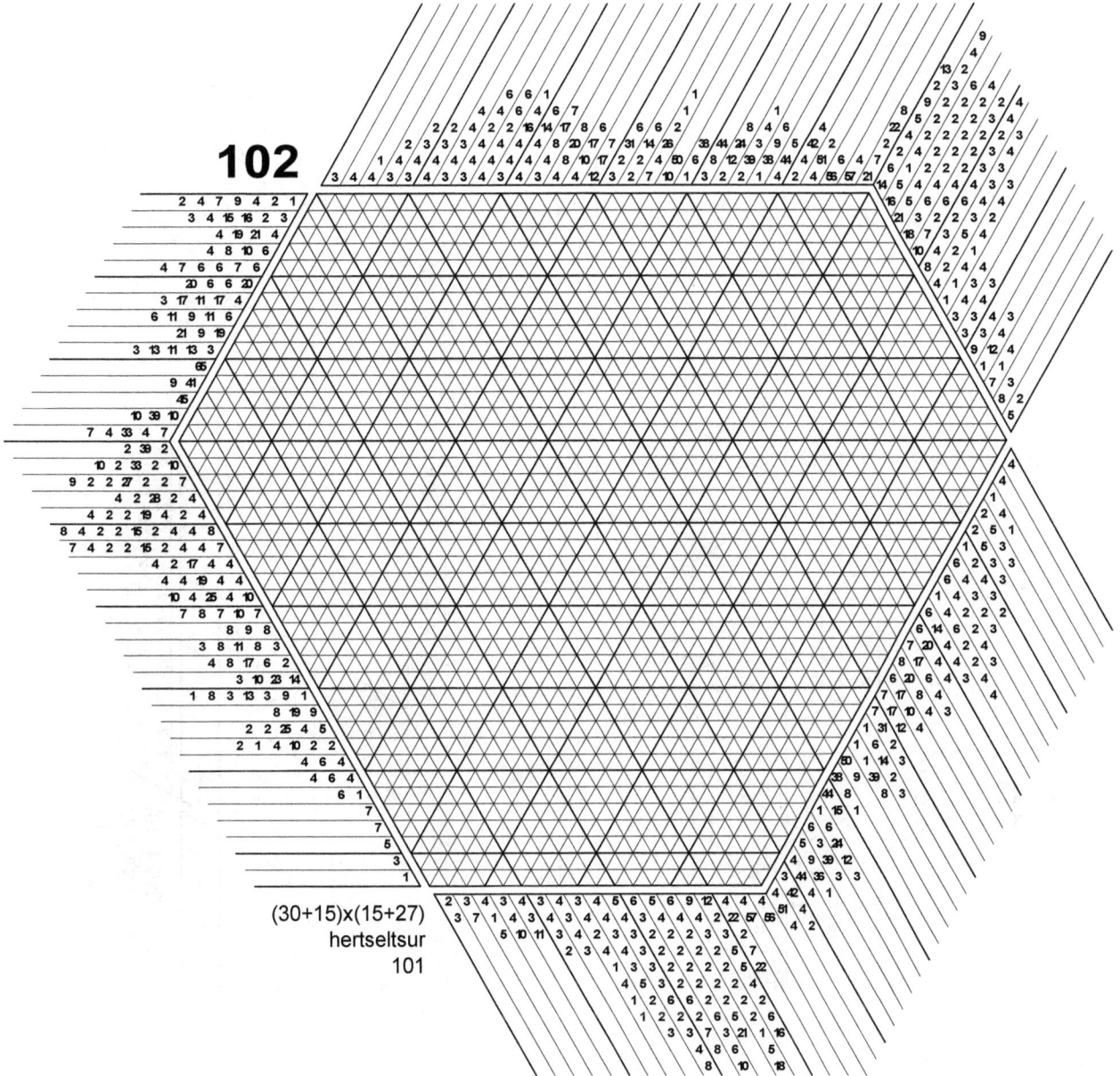

103

(26+12)x(24+18)
ANA_LILI
_102

104

(16+15)x(15+16)
ANA_LILI
102

105

(5+5)x(5+5)
Sliam
102

106

(5+5)x(5+5)
zjmonty
102

107

(18+15)x(7+25)
ANA_LILI
102

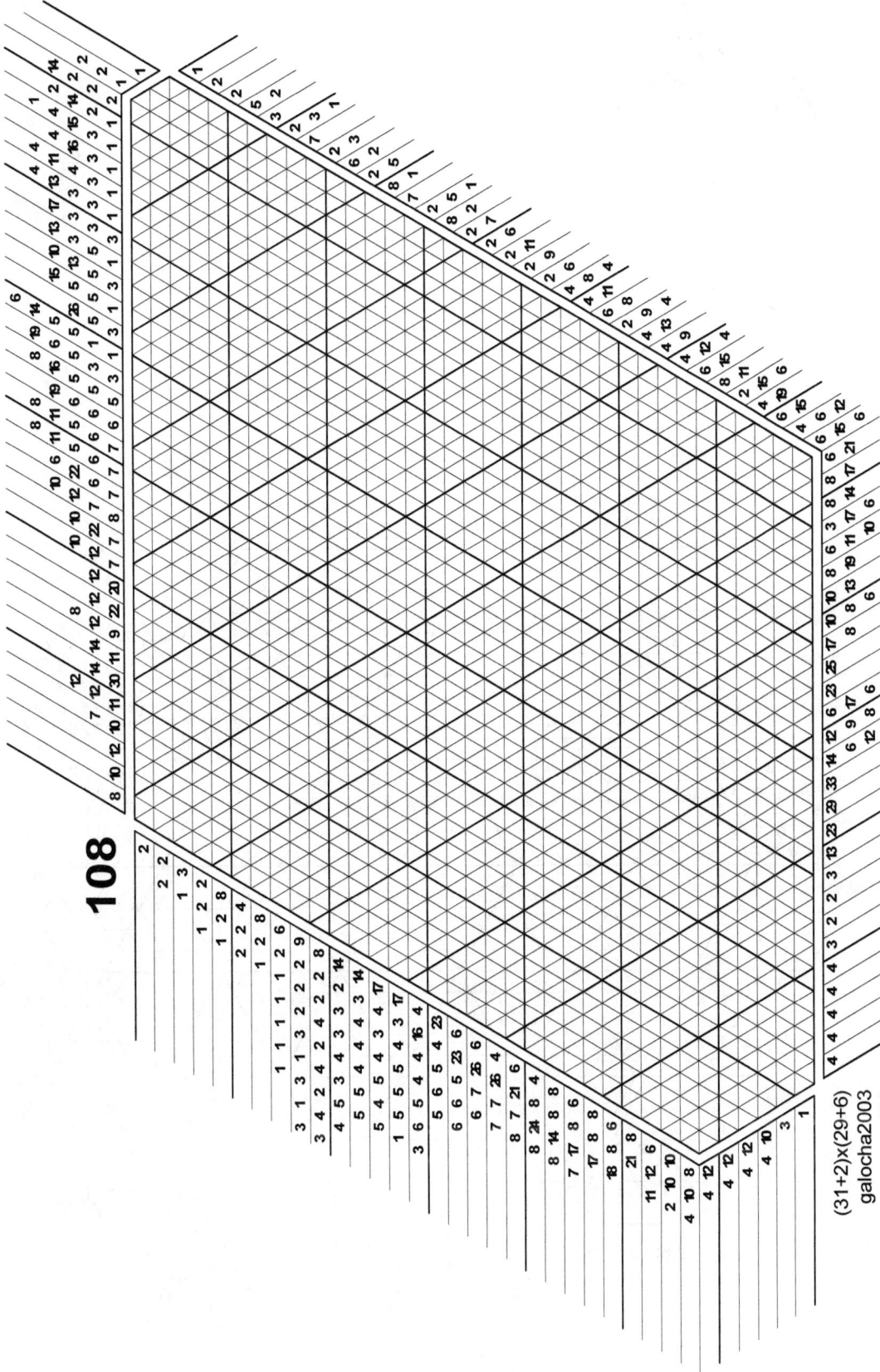

108

(31+2)x(29+6)
galocha2003
102

109

(5+5)x(5+5)
zjmonty
102

110

(14+17)x(15+17)
ANA_LILI
102

111

(12+9)x(4+11)
Vargflickan
102

112

(16+18)x(20+8)
ANA_LILI
102

113

(29+15)x(20+15)
Lenchik
102

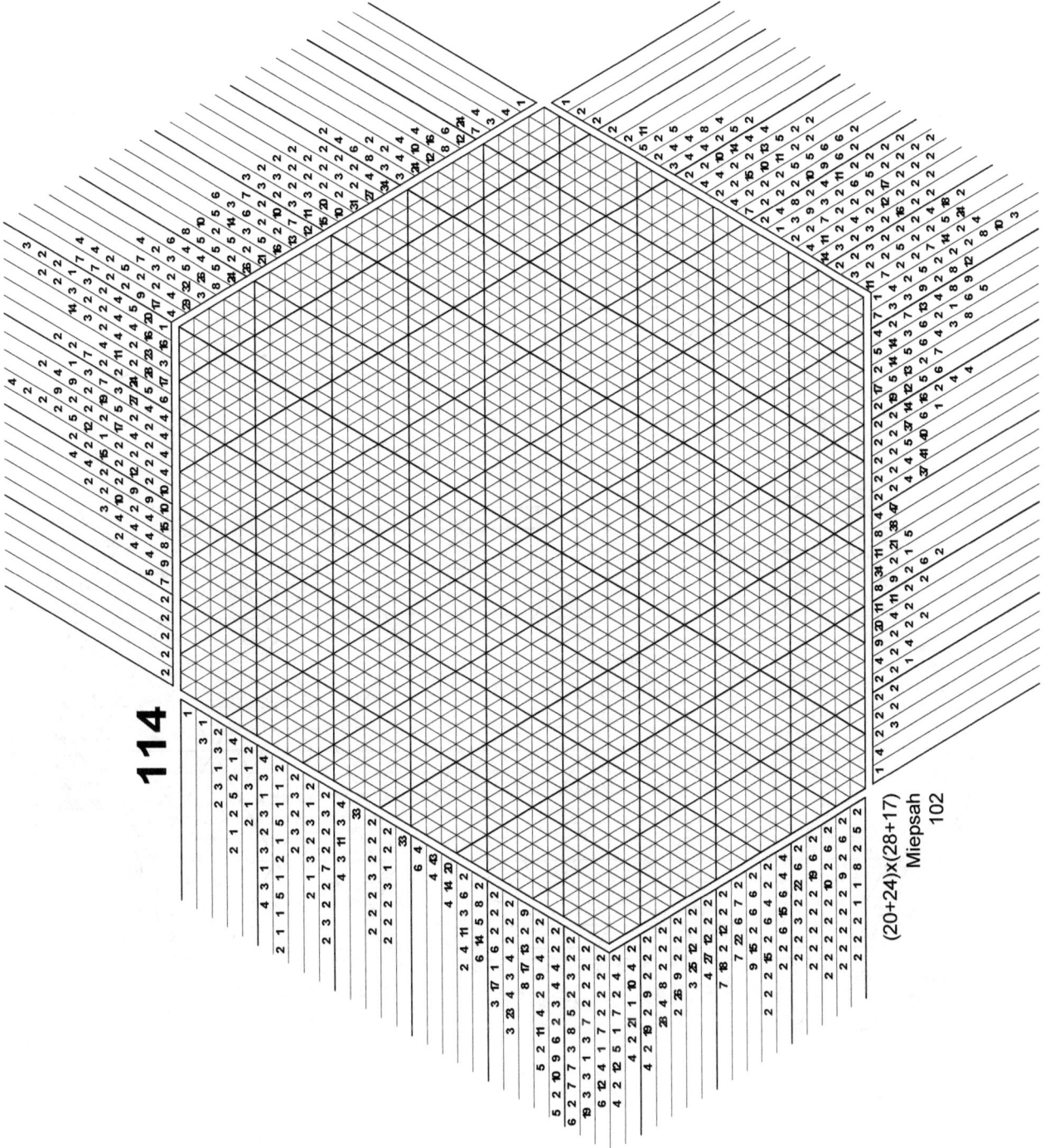

114

(20+24)x(28+17)
Miepsah
102

115

(13+22)x(25+9)
ANA_LILI
102

116

(23+9)x(4+9)
amsterdam
102

117

(16+15)x(19+14)
ANA_LILI
102

118

(4+24)x(34+0)
ANA_LILI
102

119

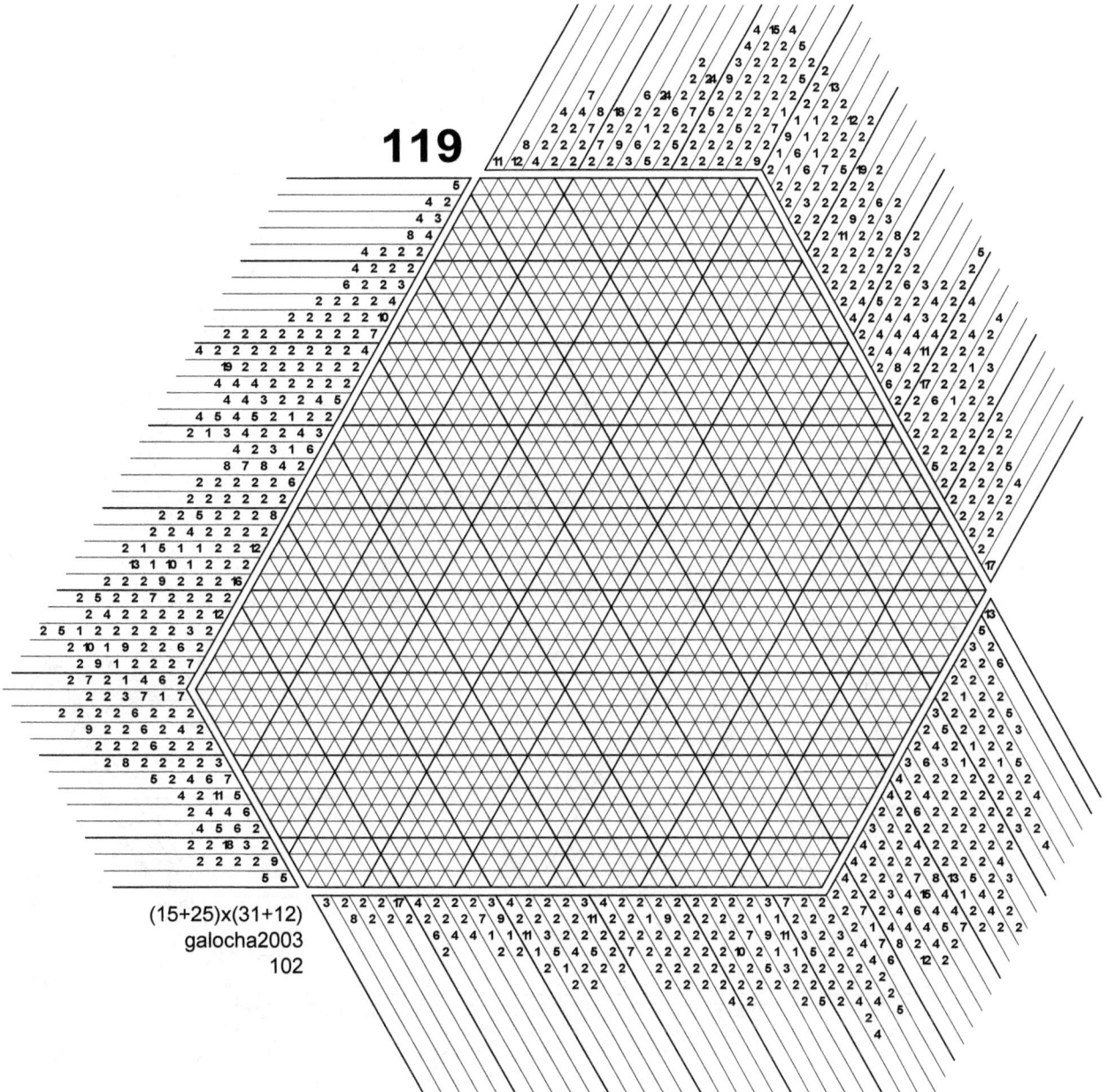

(15+25)x(31+12)
galocha2003
102

www.griddlers.net

120

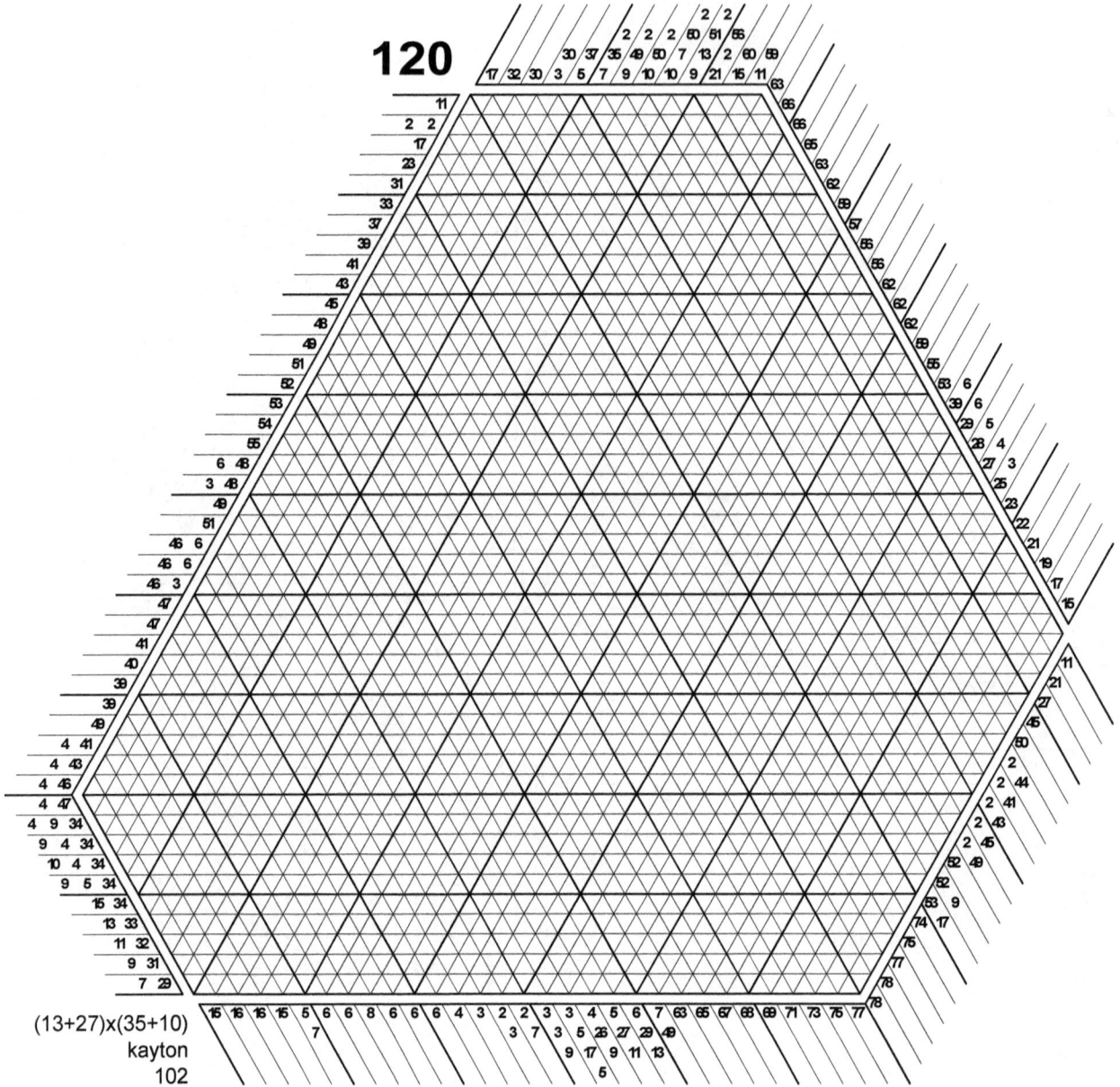

(13+27)x(35+10)
kayton
102

121

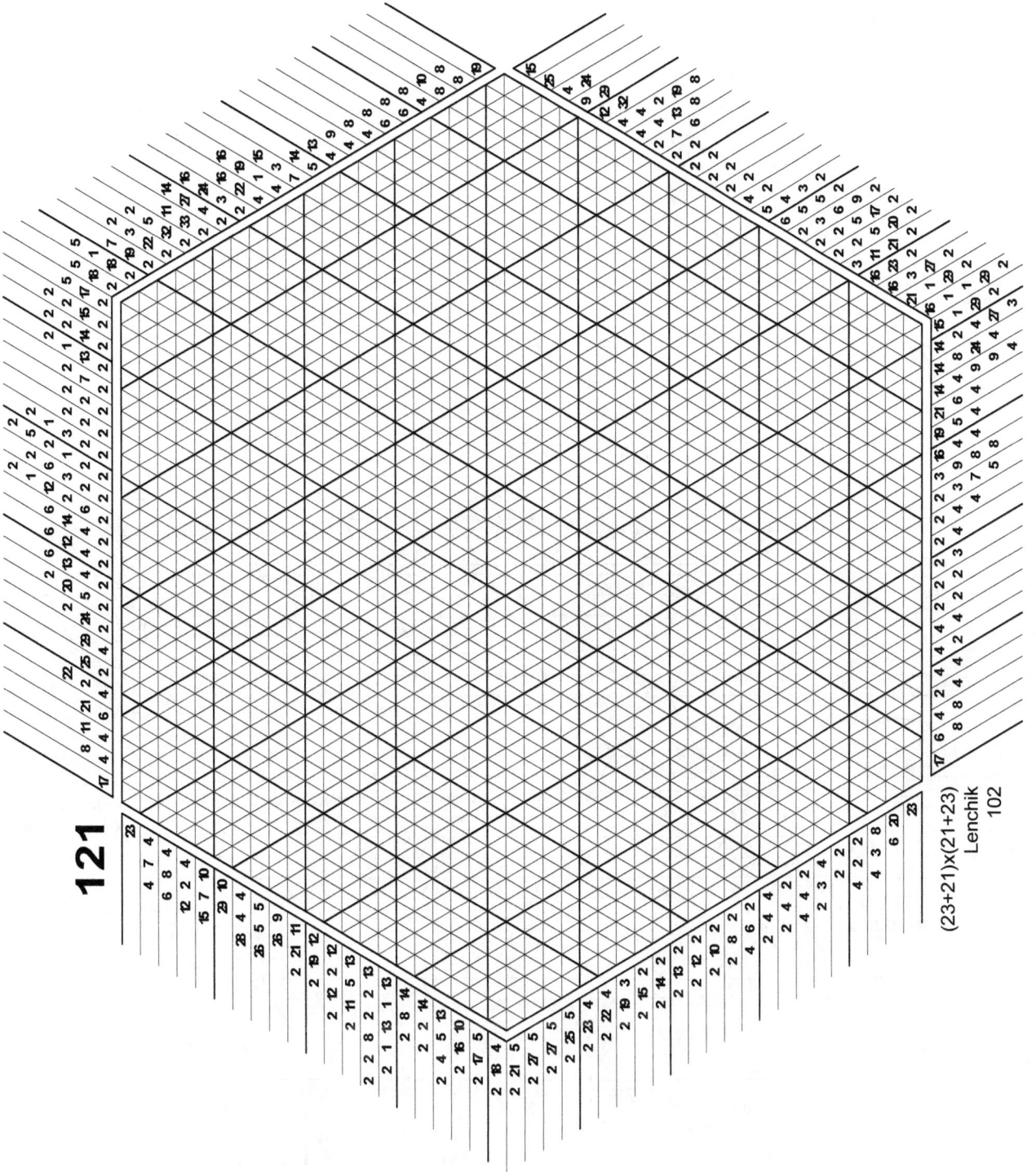

(23+21)x(21+23)
Lenchik
102

122

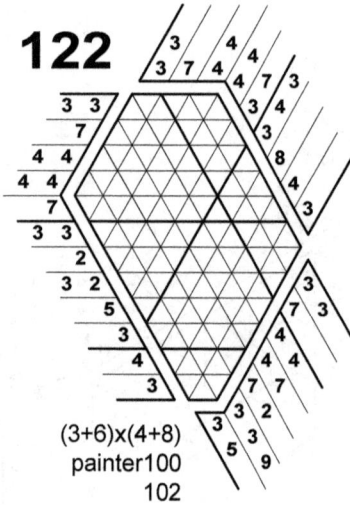

(3+6)x(4+8)
painter100
102

123

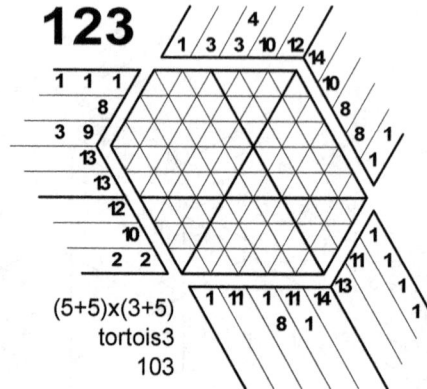

(5+5)x(3+5)
tortois3
103

124

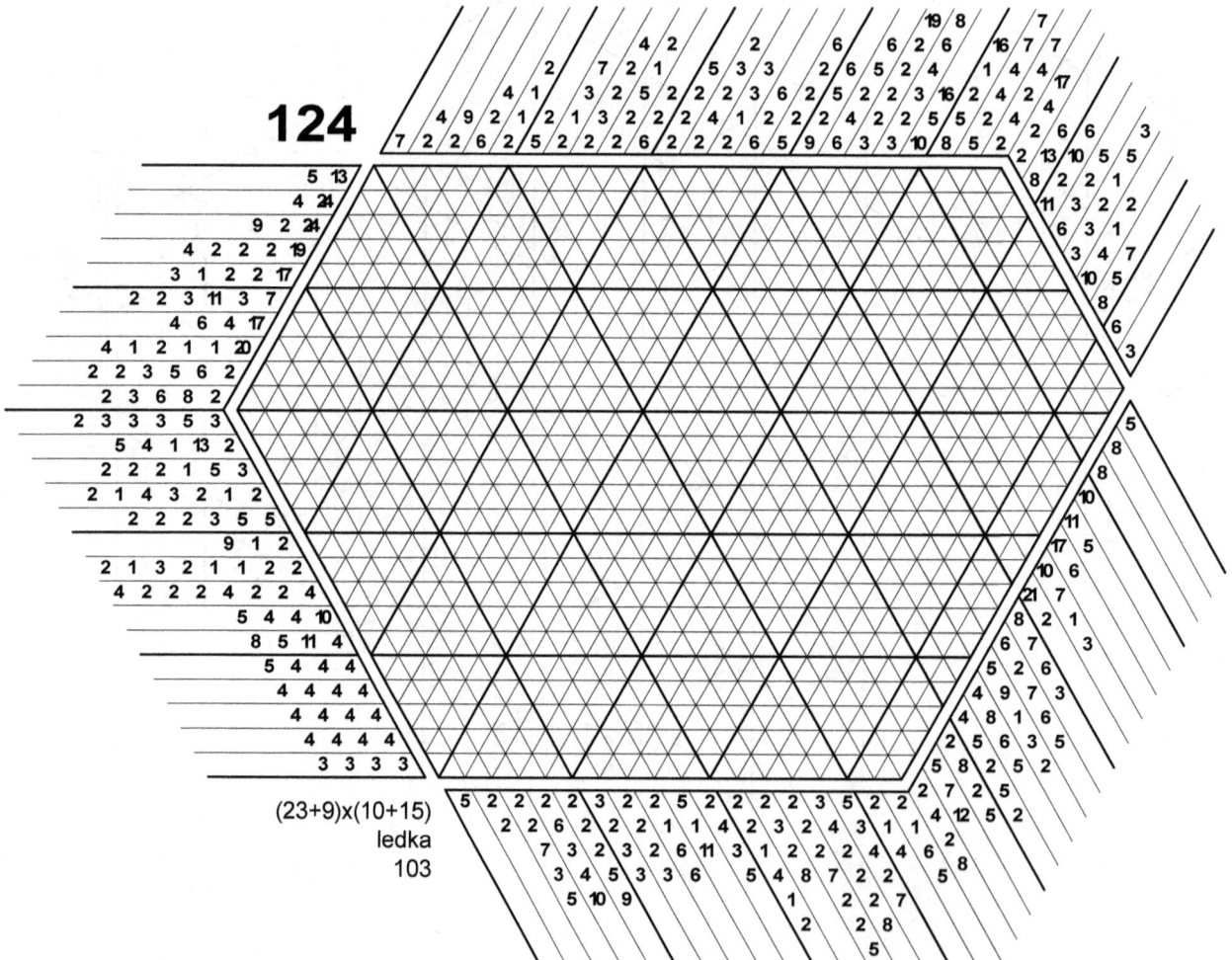

(23+9)x(10+15)
ledka
103

125

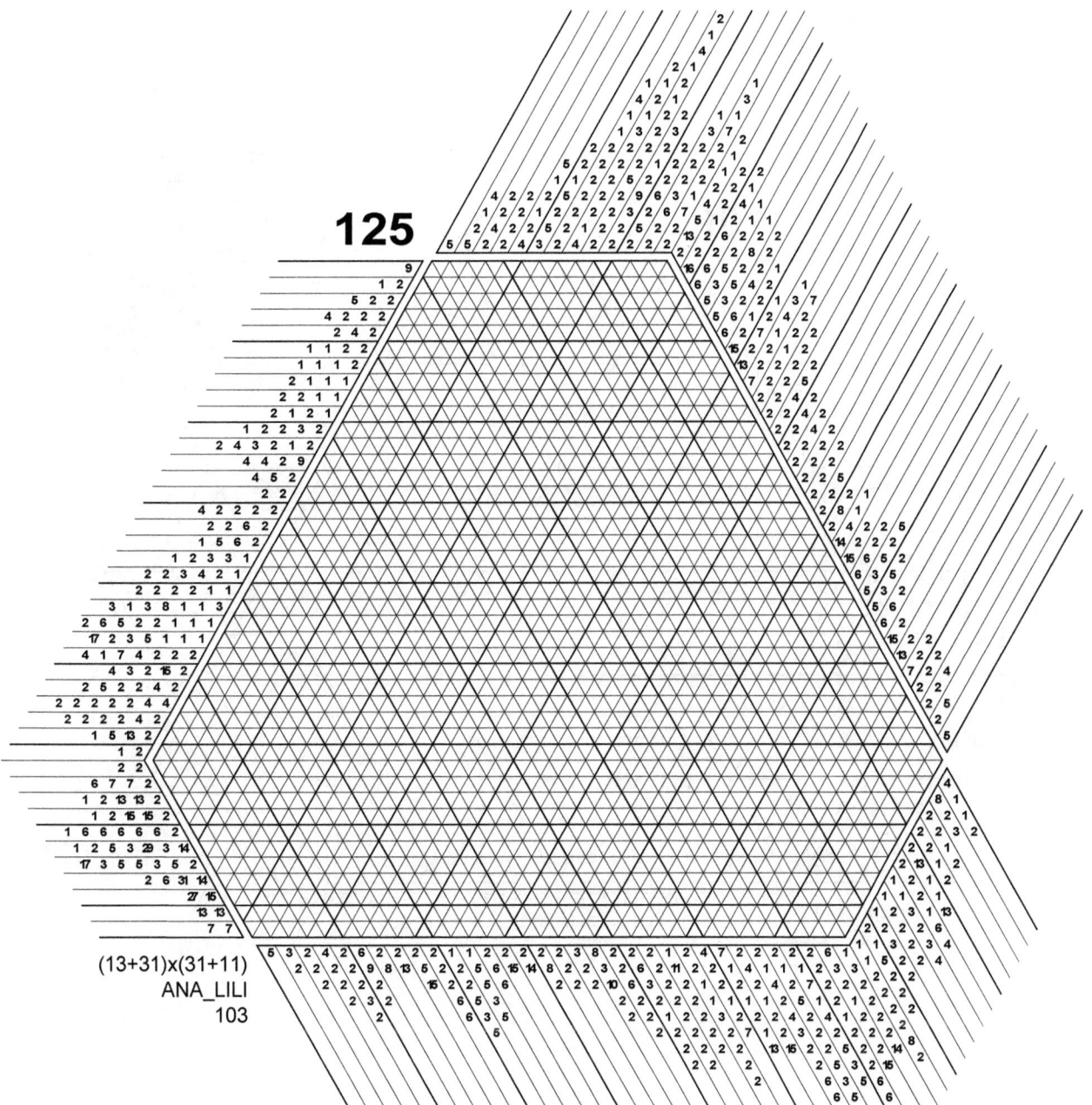

(13+31)x(31+11)
ANA_LILI
103

126

(20+12)x(20+10)
ANA_LILI
103

127

(18+23)x(18+22)
ANA_LILI
103

128

(15+24)x(35+10)
hertseltsur
103

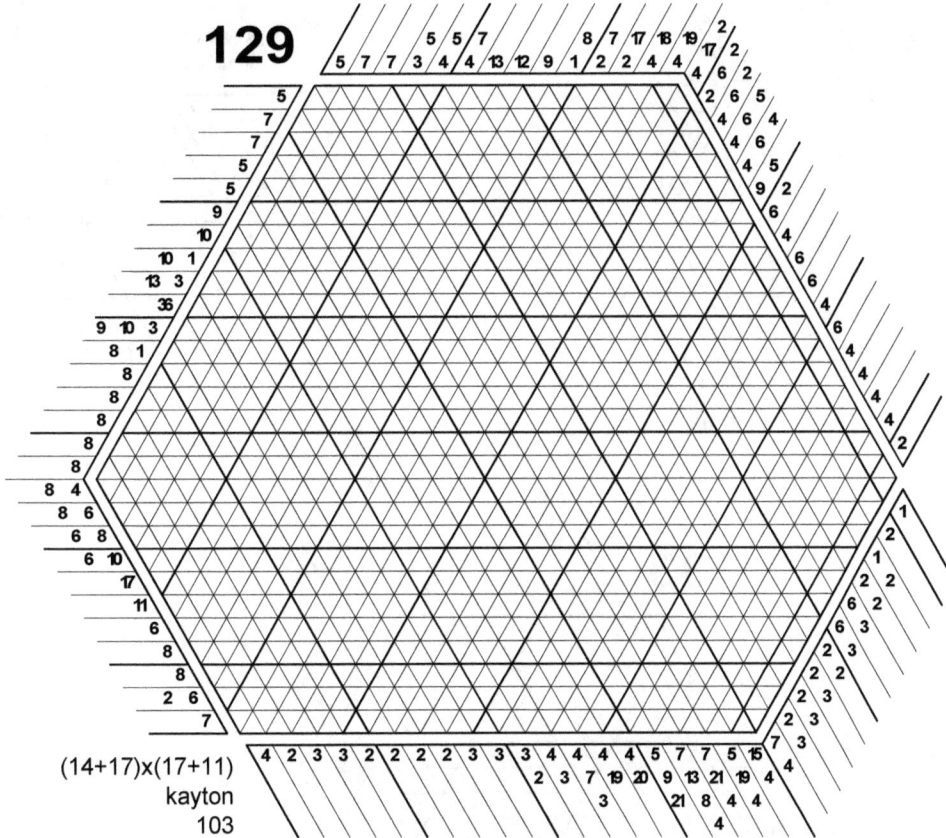

129

(14+17)x(17+11)
kayton
103

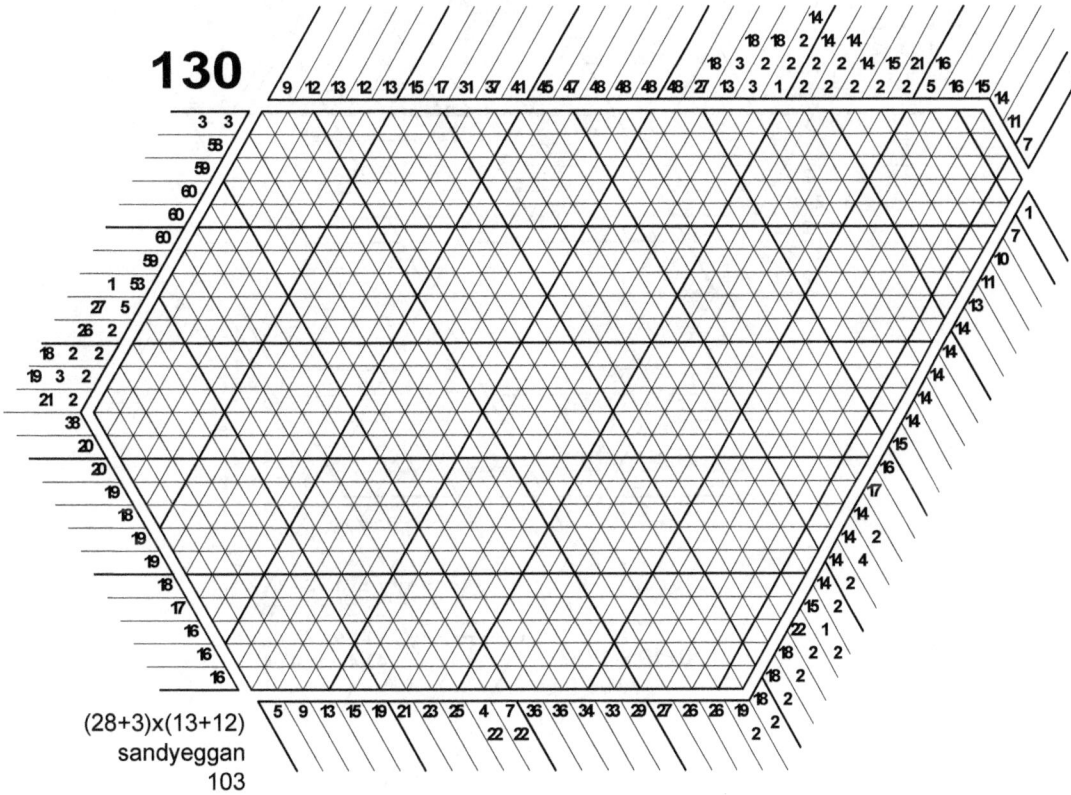

130

(28+3)x(13+12)
sandyeggan
103

131

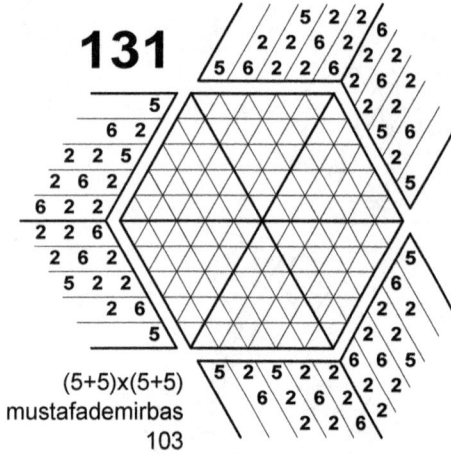

(5+5)x(5+5)
mustafademirbas
103

132

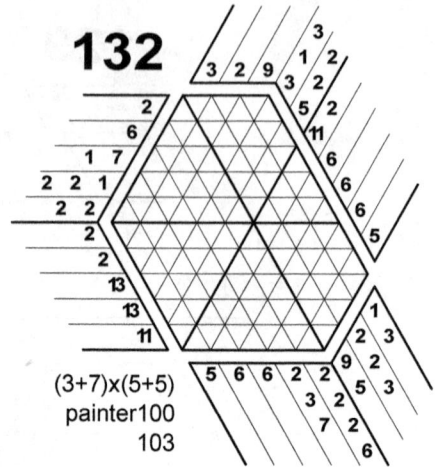

(3+7)x(5+5)
painter100
103

133

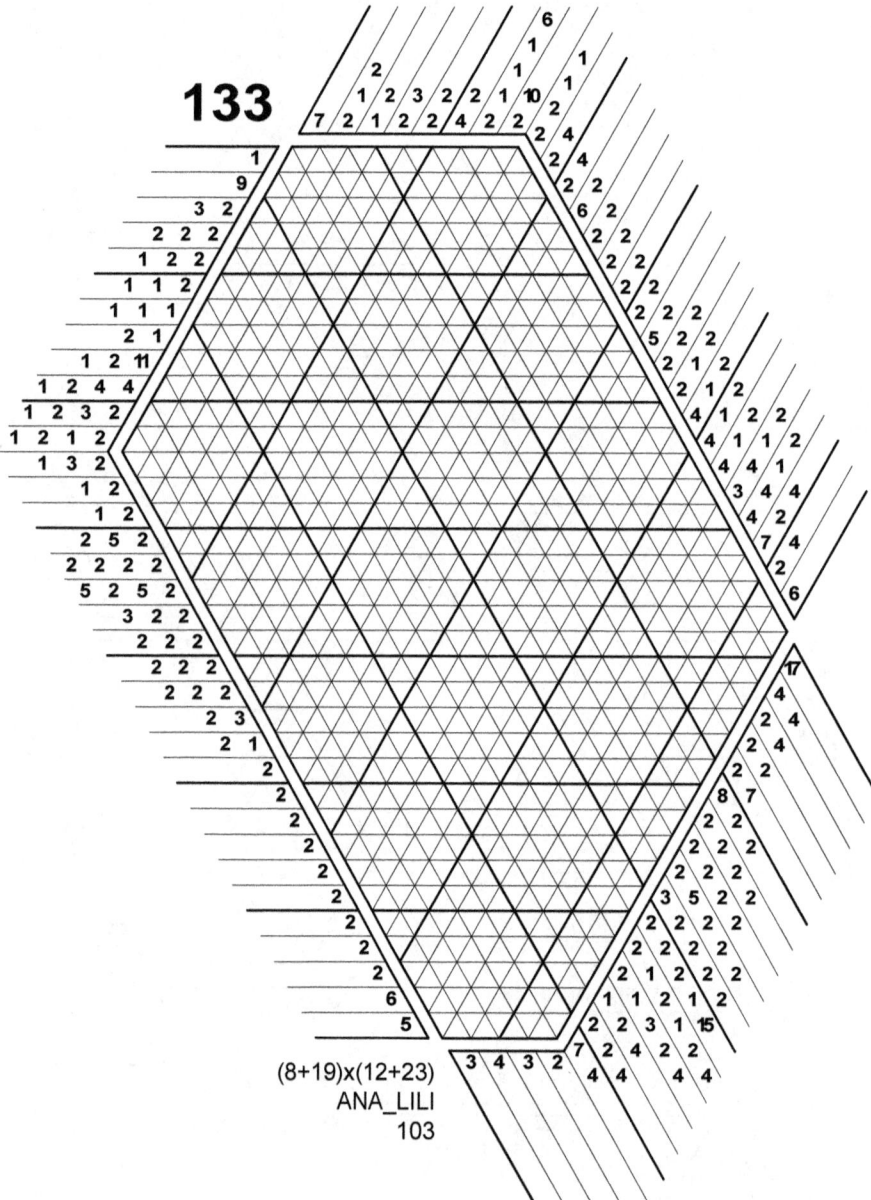

(8+19)x(12+23)
ANA_LILI
103

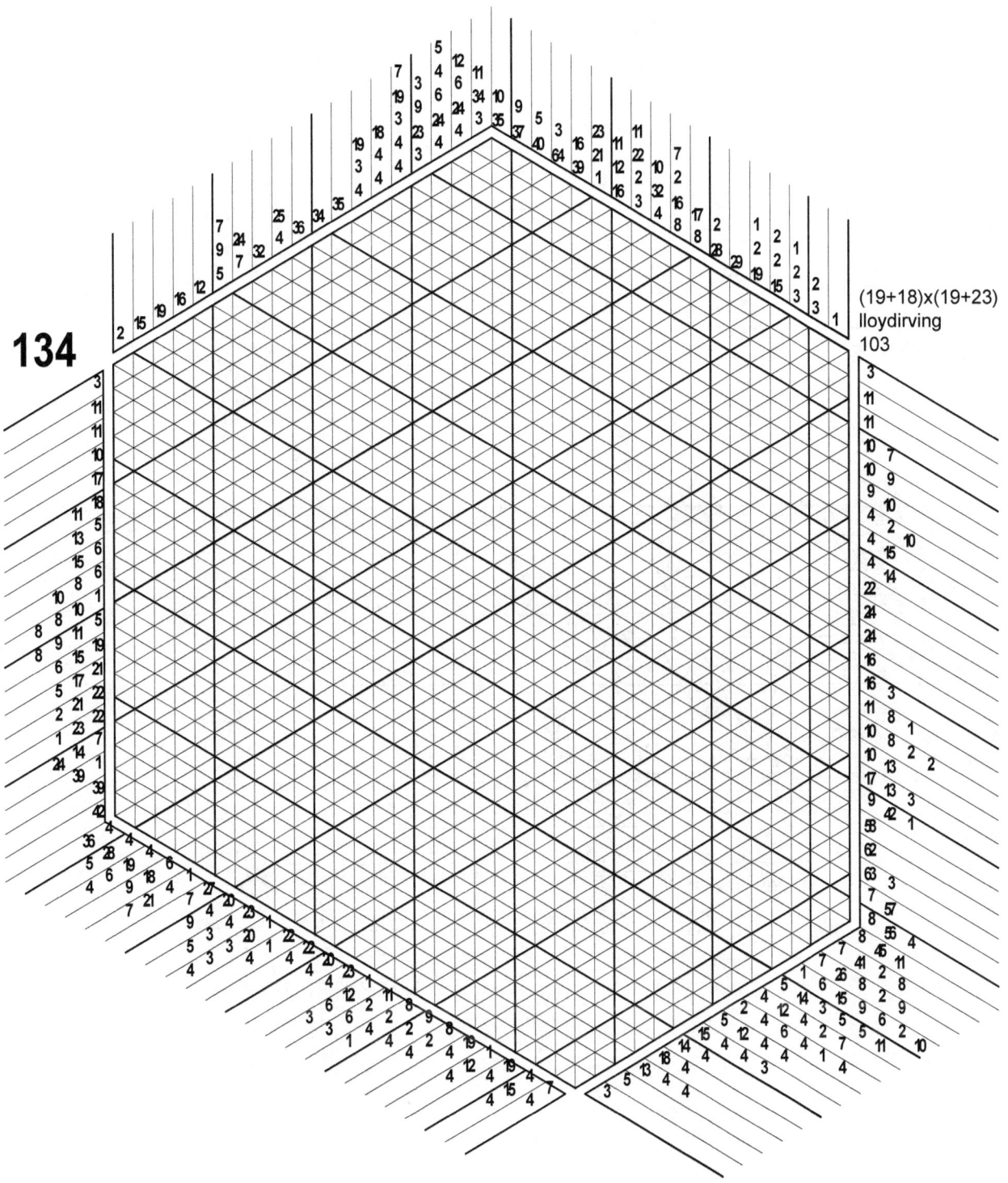

134

(19+18)x(19+23)
lloydirving
103

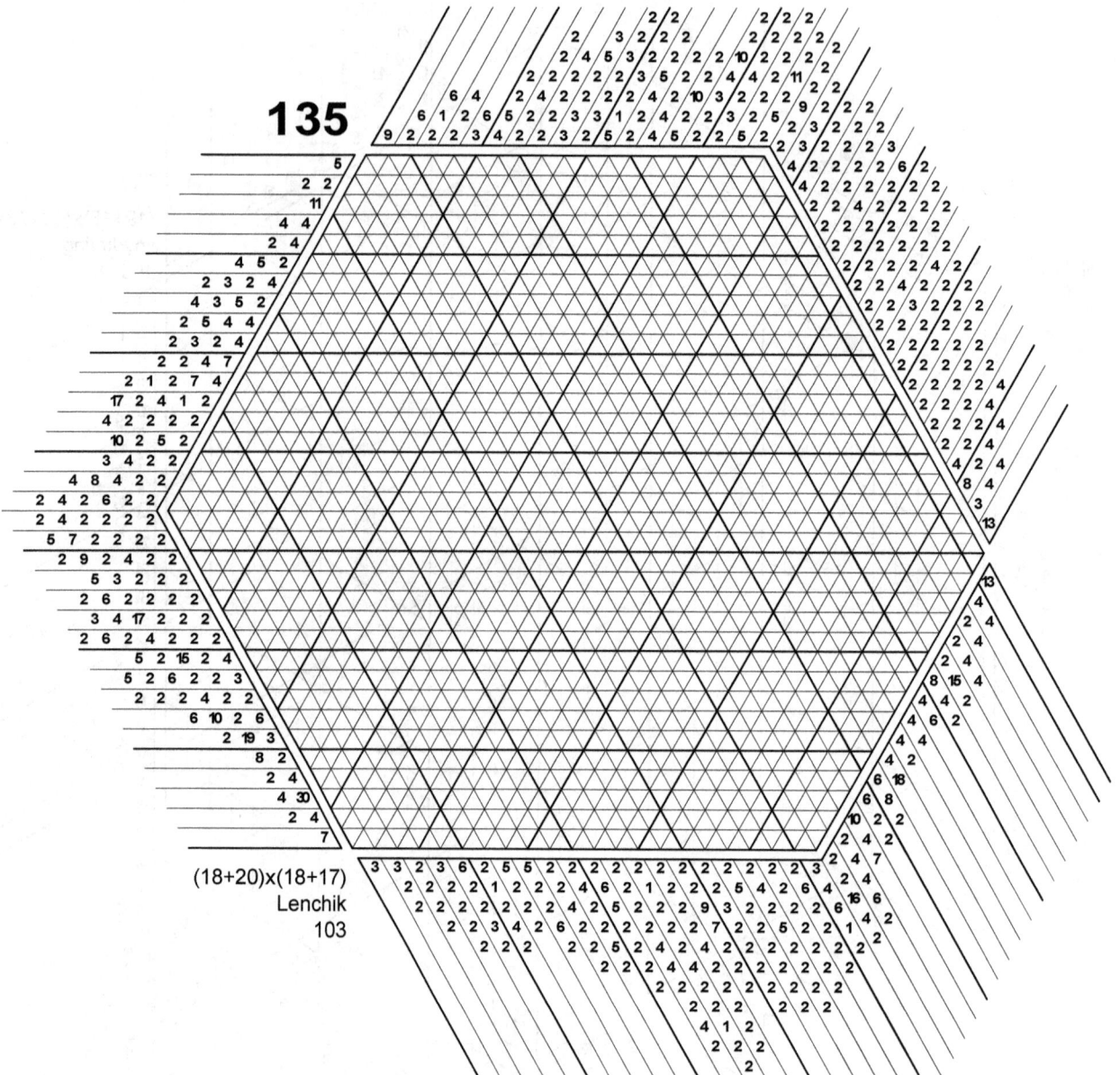

135

(18+20)x(18+17)
Lenchik
103

136

(15+22)x(32+11)
Lenchik
103

137

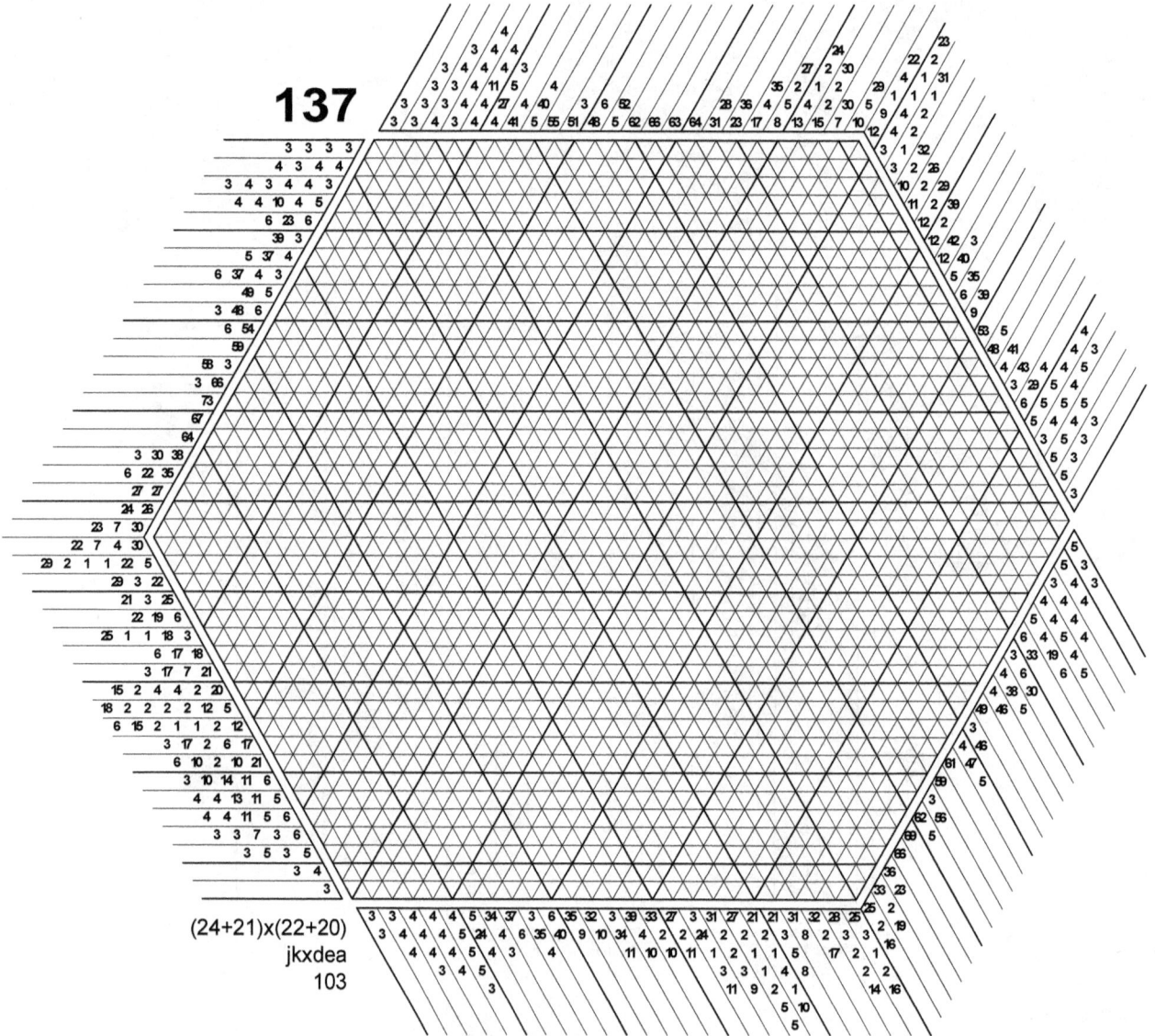

(24+21)x(22+20)
jkxdea
103

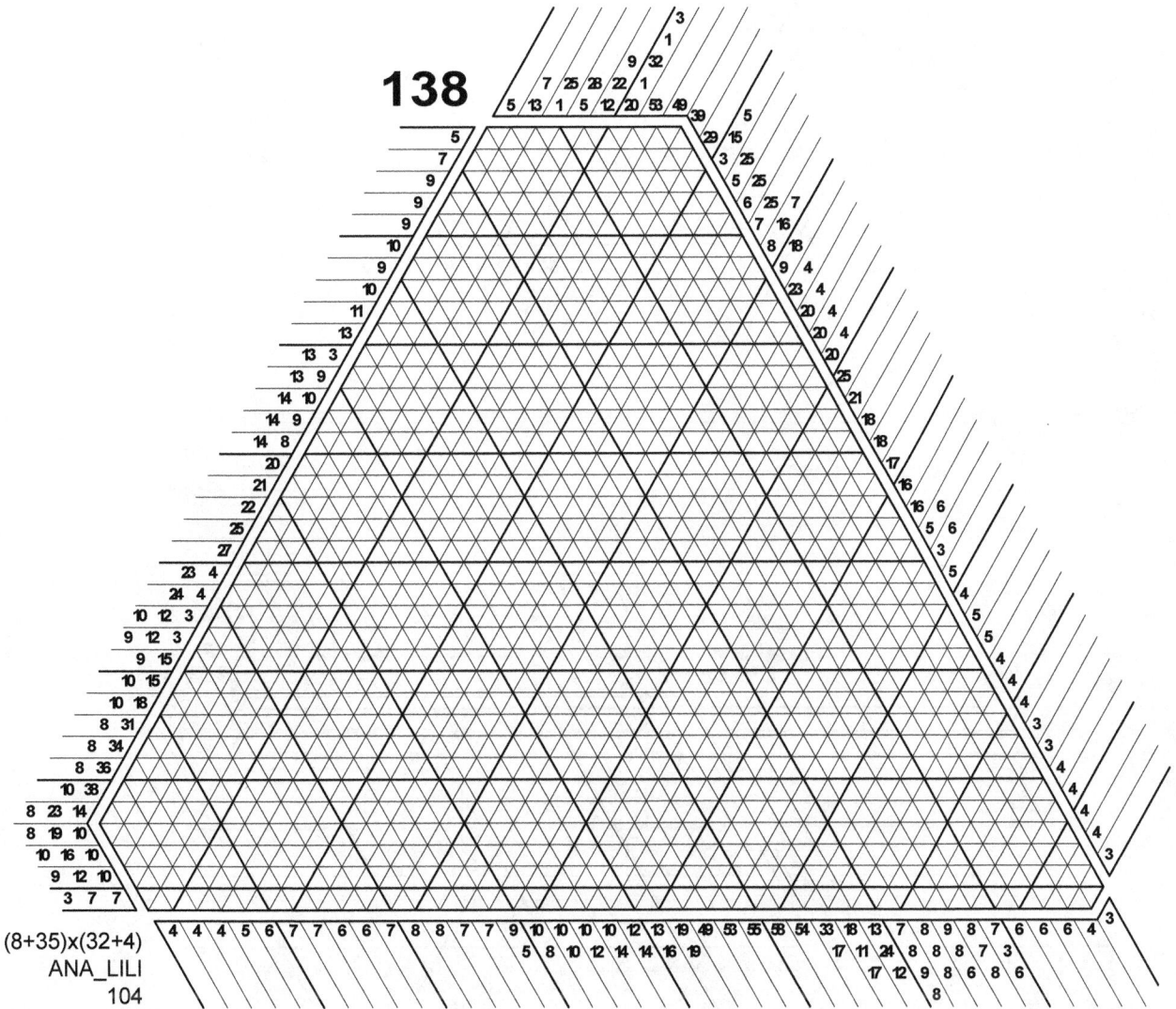

138

(8+35)x(32+4)
ANA_LILI
104

139

(18+7)x(7+7)
shadow2097
104

140

(16+9)x(11+2)
bilbobilbo
104

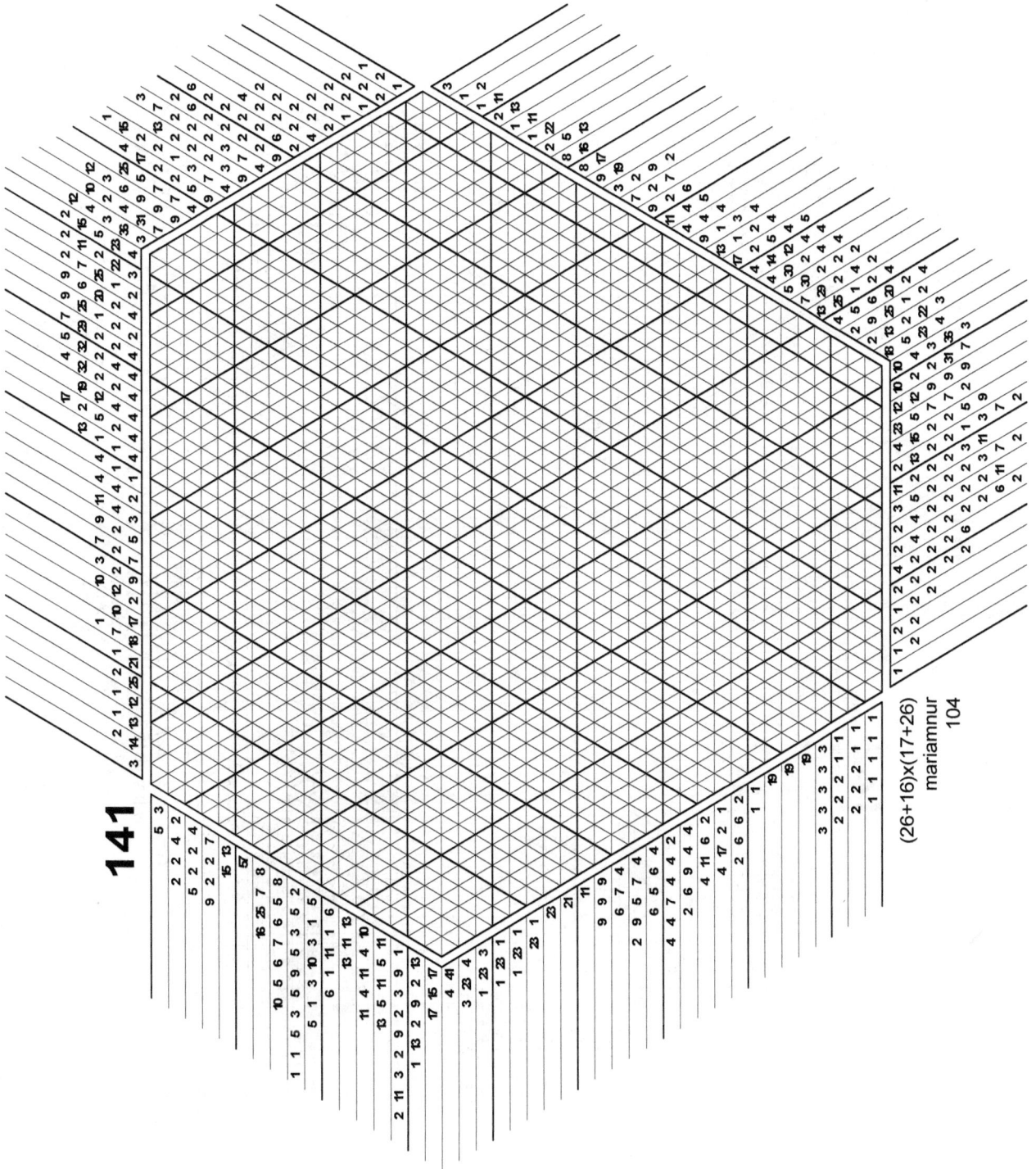

141

(26+16)x(17+26)
mariamnur
104

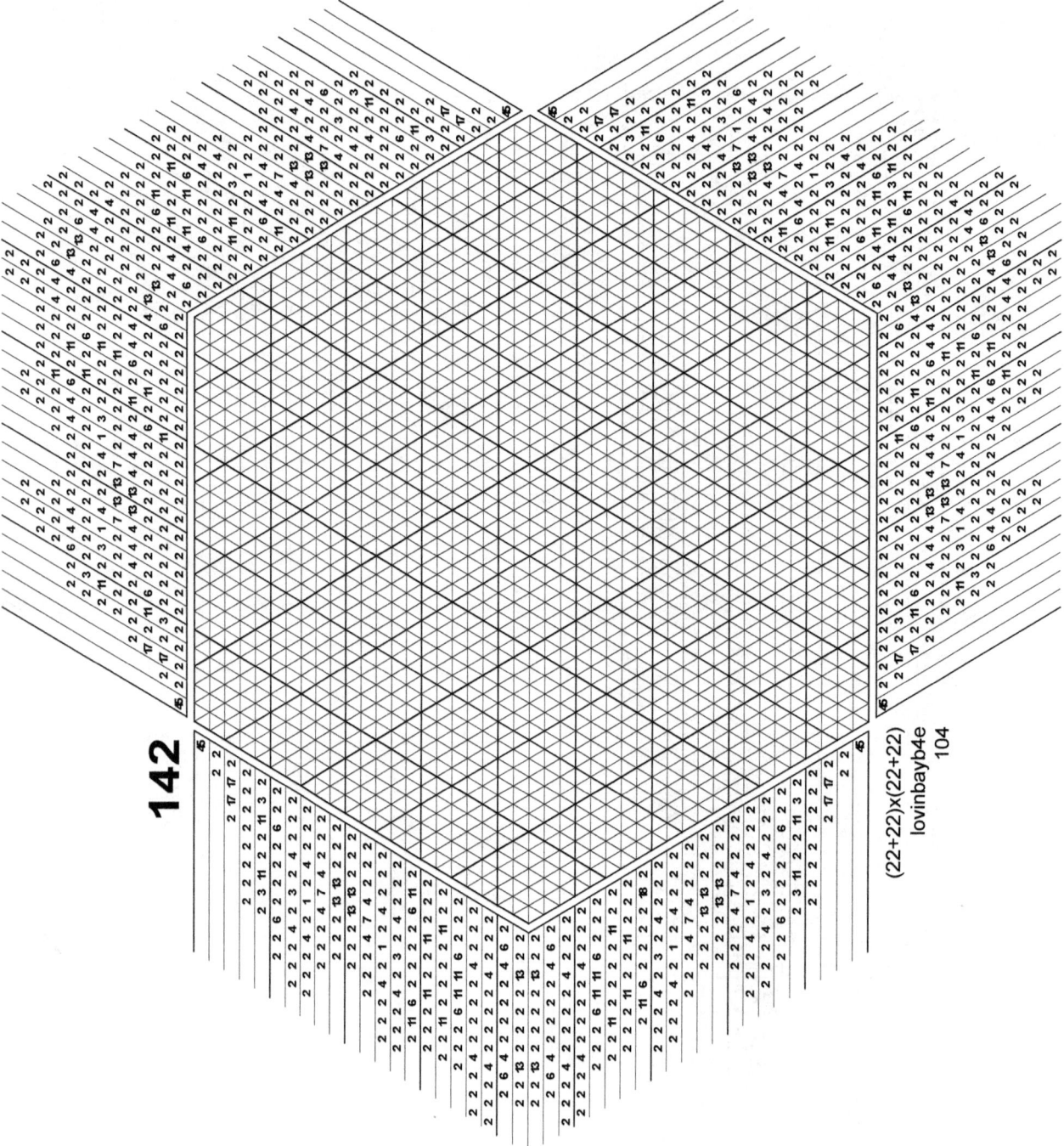

142

(22+22)x(22+22)
lovinbayb4e
104

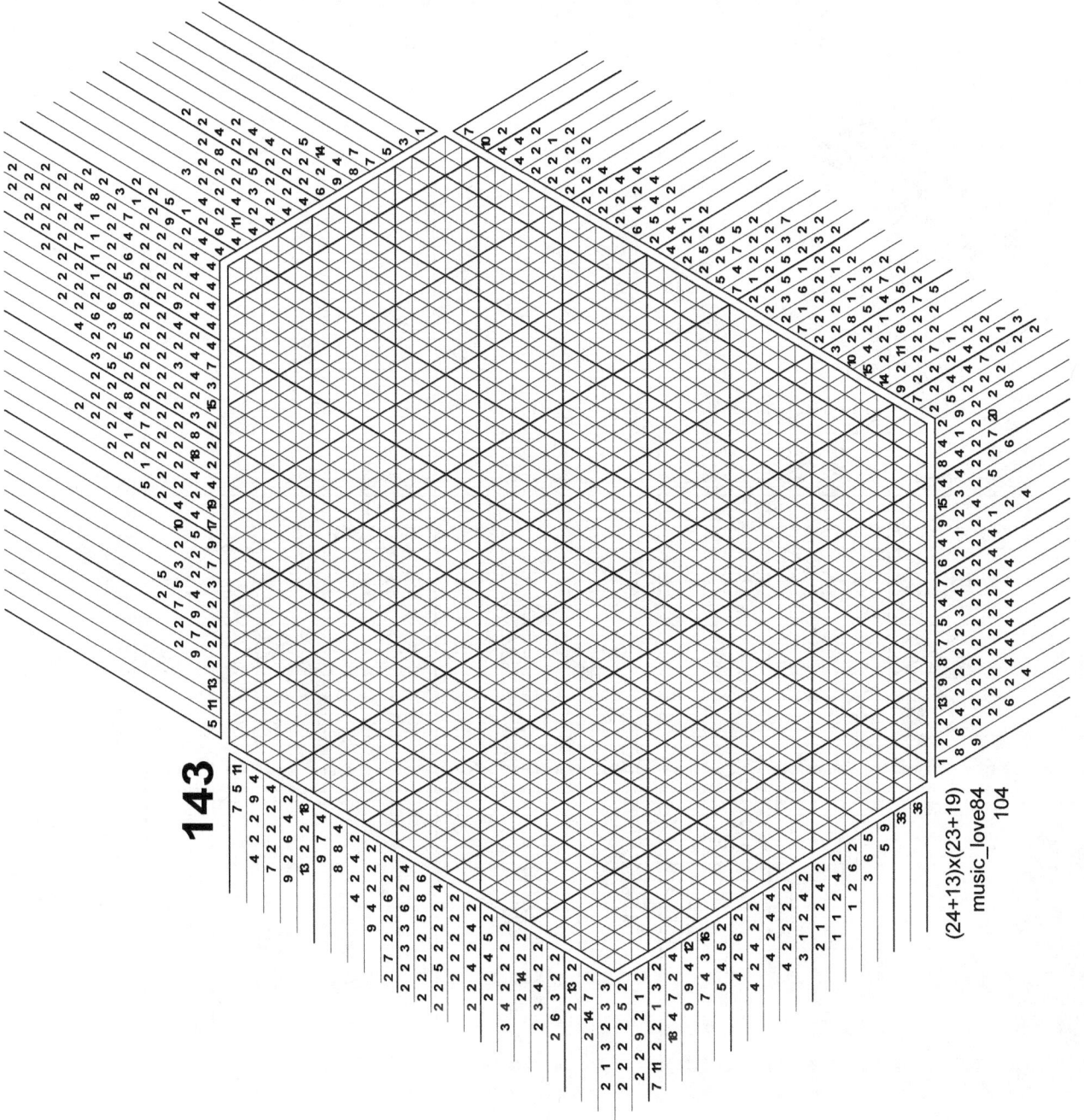

143

(24+13)×(23+19)
music_love84
104

144

(18+16)x(17+16)
jadespade
104

145

(10+12)x(13+1)
Teeaz
104

146

(14+14)x(17+14)
ledka
104

147

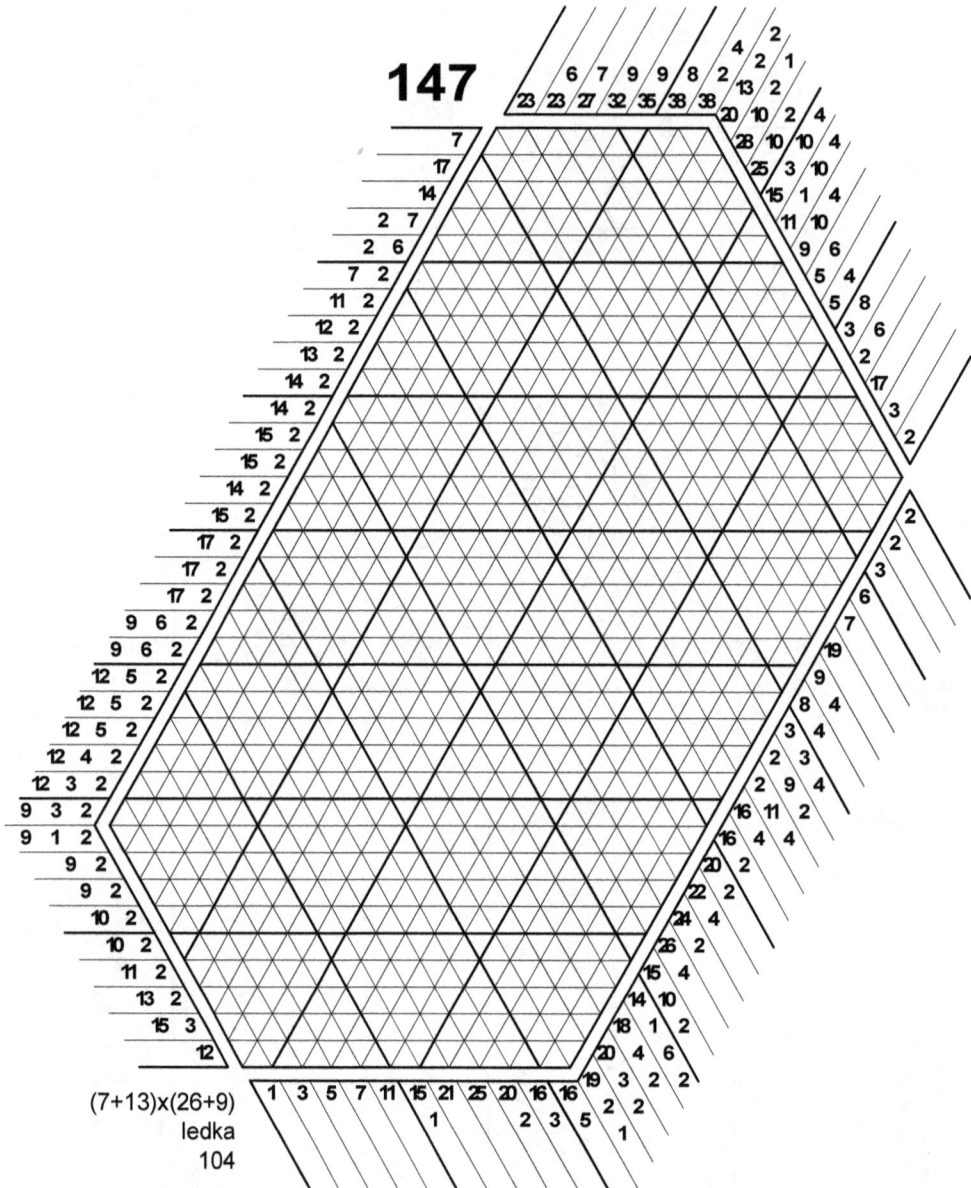

(7+13)x(26+9)
ledka
104

148

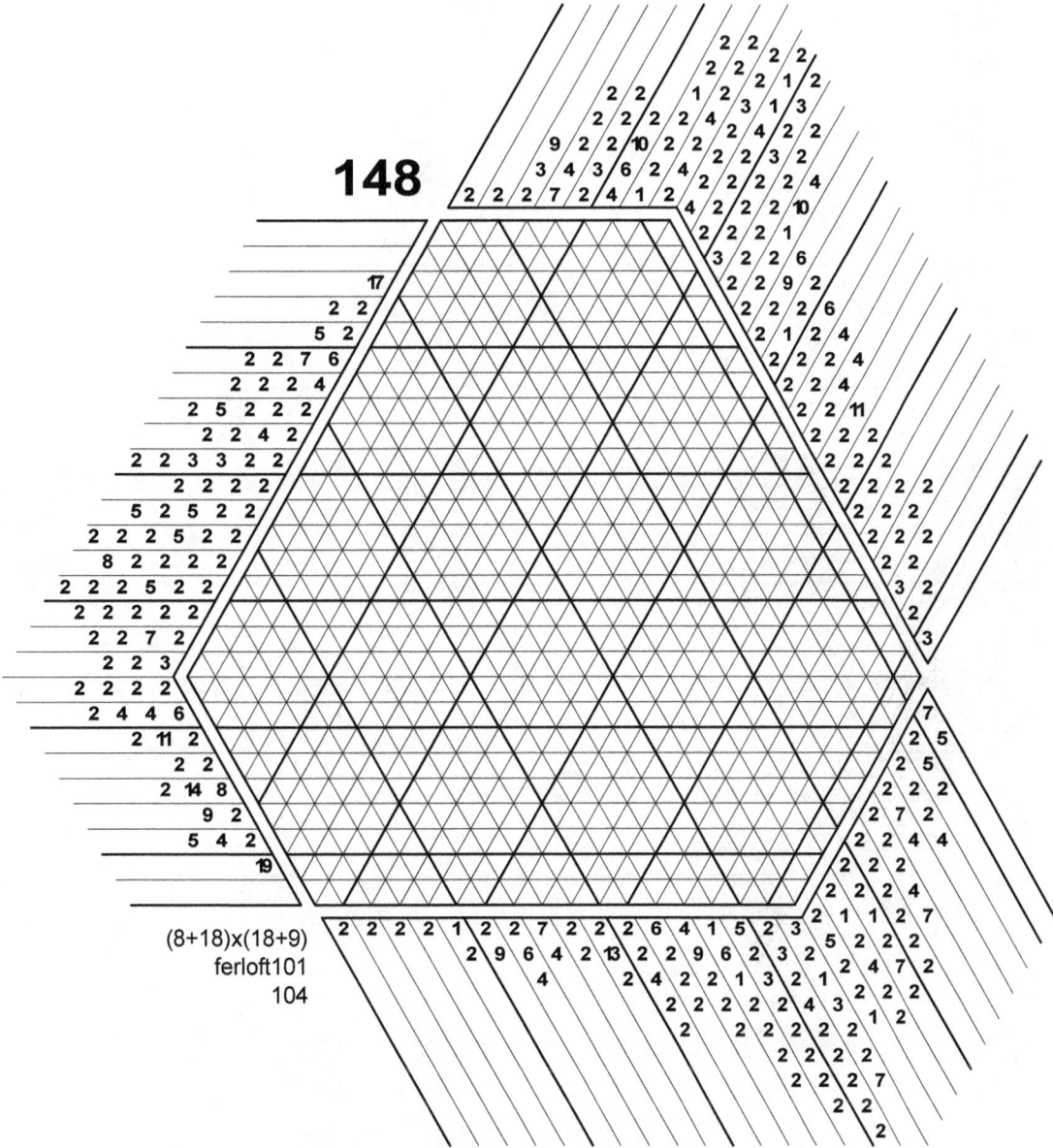

(8+18)x(18+9)
ferloft101
104

149

(13+23)x(21+24)
Lenchik
104

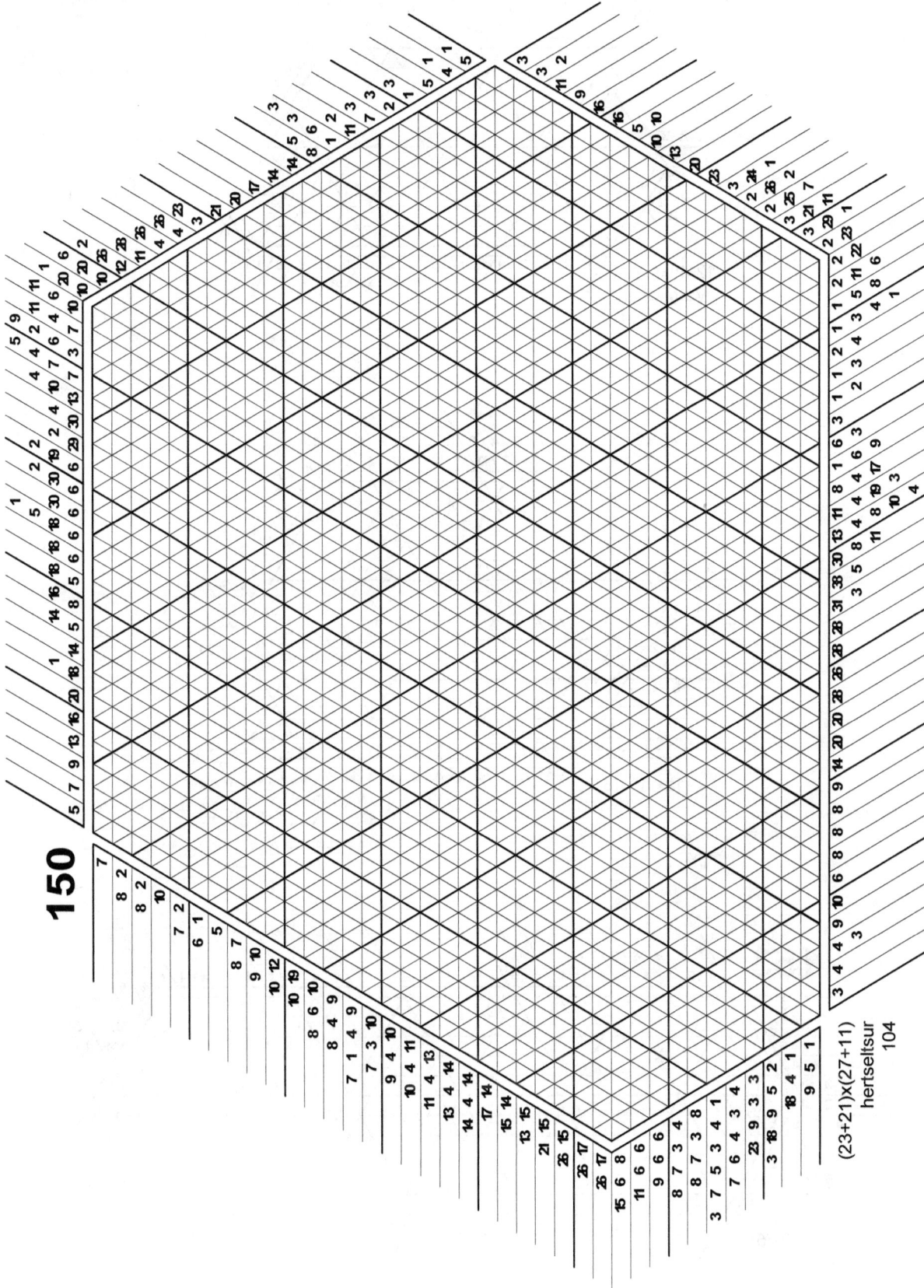

150

(23+21)x(27+11)
hertseltsur
104

151

(12+8)x(6+4)
Vargflickan
104

152

(12+10)x(6+8)
Ra100
104

153

(26+7)x(5+6)
jadespade
104

154

(35+10)x(10+22)
mariamnur
104

155

(12+14)x(14+14)
podjedzona
104

156

(9+20)x(15+14)
Lenchik
104

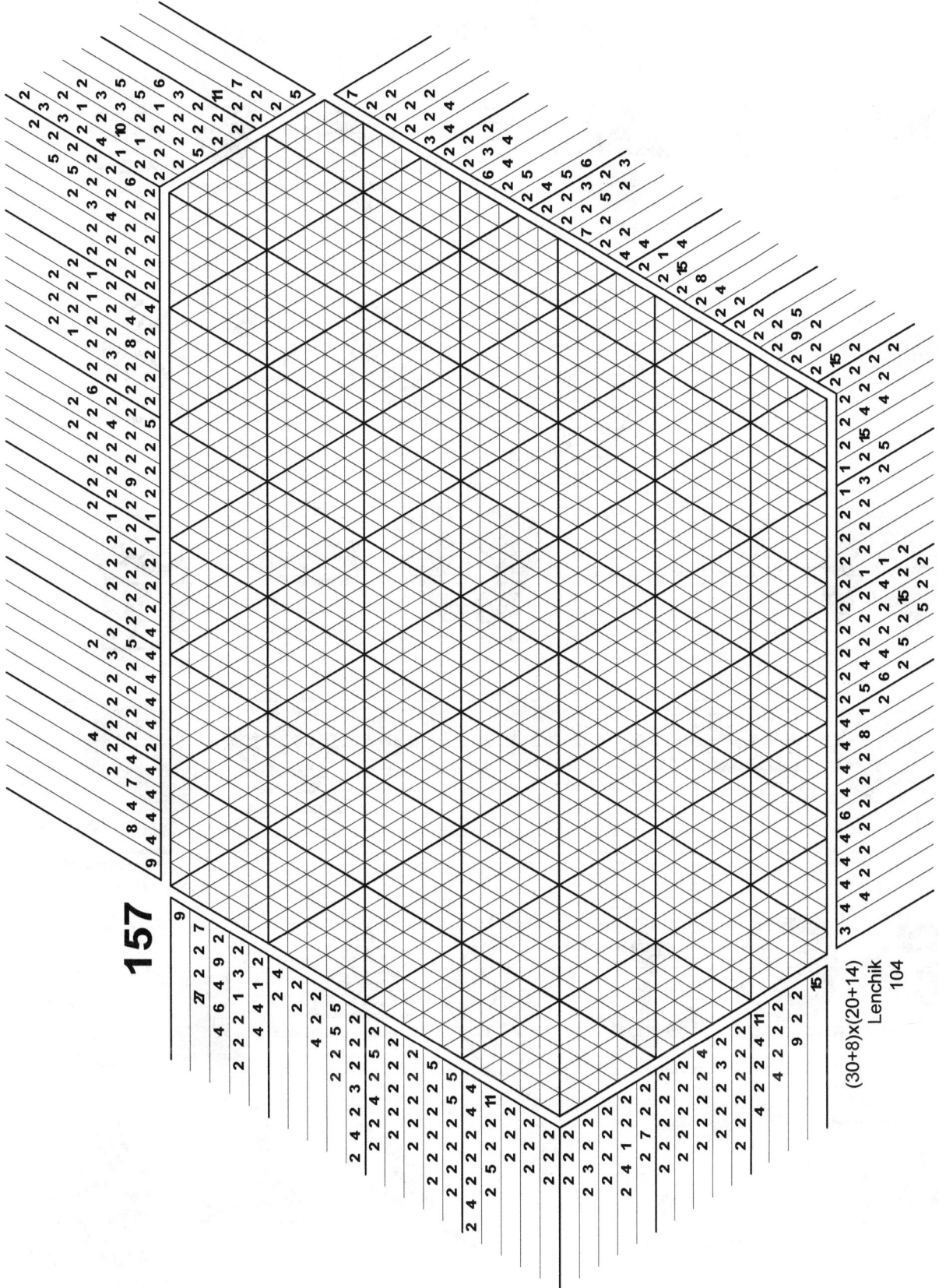

157

Lenchik
104

Solutions

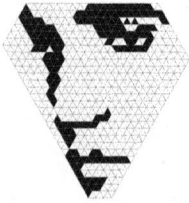

1: Elvis Presley

2: Kiddlers: Hiker

3: Night Prowler

4: Cat and Mouse

5: Karate Fighter

6: Chameleon

7: Panda

8: Simplified Snail

9: Kiddlers: Hold My Hand

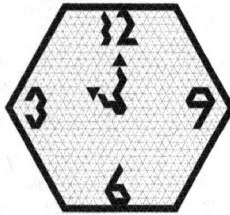

10: 2 O'clock

11: Flower

12: Table Lamp

13: Dragonfly

14: Night: Wanderer

15: Kiddlers: Kicker

16: Tree

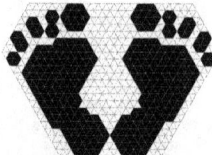

17: Footprint

18: Snowy Night

20: Black Mage

19: Parrot

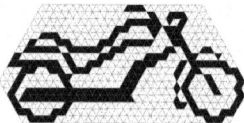

21: Bike

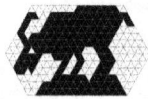

22: Goat

23: Elephant

24: Gestalt's Cube

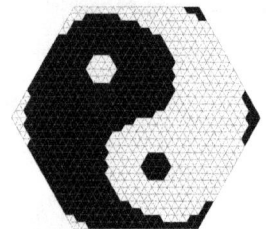

25: Yin and Yang

26: Recycling

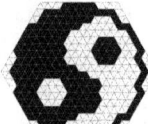

27: Yin and Yang

28: Anchor

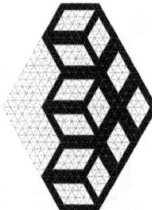

29: Cubes

30: Triangular Celtic Knot

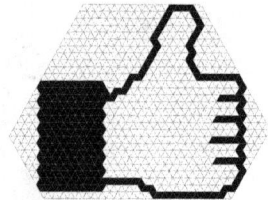

31: I Like It!

32: Scale

33: Barclays Bank Logo

34: Bee

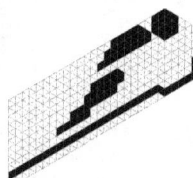

35: Ski Jump

36: Ready?

37: Rabbit

38: Couple Kissing

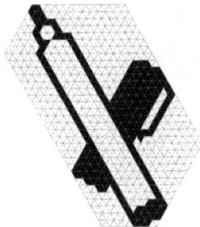

39: Pen

40: Adidas

41: Ear

42: WWF

43: A Flower for Hagit. Thank you

44: Spider

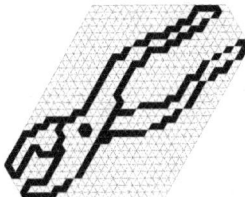

45: Pliers

46: Detroit Red Wings

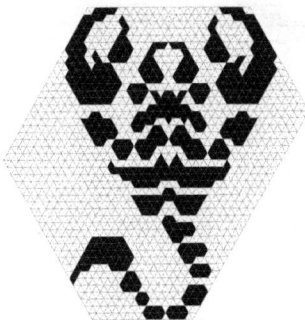

47: Scorpio Tattoo

48: Elk

49: Kiddlers: Mountain Climber

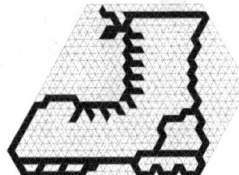

50: Boot

51: Dance in the Dark

52: Kiddlers: On One Leg

53: Scorpio

54: Bob The Builder

55: Polar Bear

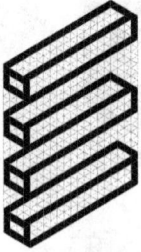
56: How Many Planks?

57: Bonsai

58: Spiderweb

59: Crab

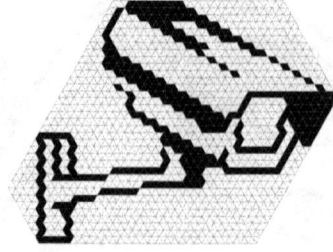
60: Camera

61: Wind Surf

62: Bear

63: Dog

64: Roller Skate

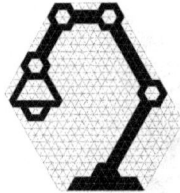
65: Lamp

66: Frog

67: Soccer

68: Skating

69: Hare

70: Wink

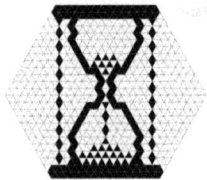
71: Hourglass

72: Creepy Tree

73: Shopping Cart

74: Safety Pin

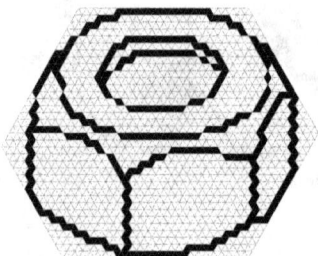
75: Nut

76: Duplicate 0. Funky Sun

77: Bat

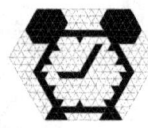
78: Alarm Clock

79: Sailboat

80: I am Not Sad, My Face Looks Like This

81: Zebra

82: Dan Fleming - Owl

83: Bells

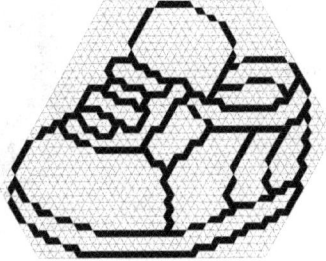

84: Bath

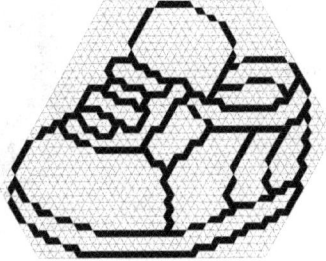

85: Shoe

86: Mushroom

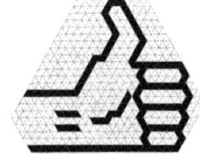

87: Thumb Up

88: Rocking Horse

89: Bull

90: Bug

91: Under Construction

92: Snow

93: Eagle

94: Magpie

95: Stormtrooper Helmet

96: Witch

97: Leaf

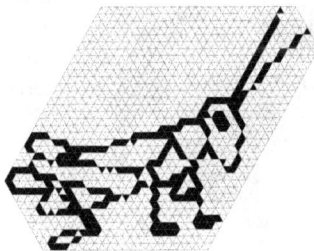

98: Rose

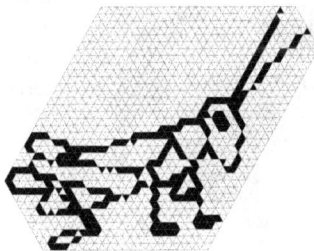

99: Grasshopper

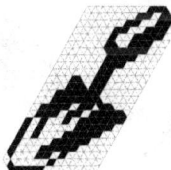

100: Spade

101: Flag

102: Albania Coat of Arms

103: Bouquet

104: Half Nectarine

105: Cube 106: Kiddlers: Quidditch

107: Discobolus

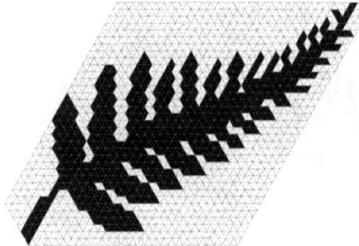

108: New Zeland Football Federation

109: Kiddlers: Taking Out the Trash

110: Felix

111: Camel

112: Scale

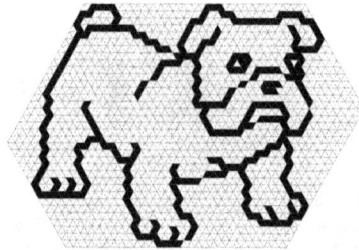

113: Bulldog

114: Royal

115: Ladies and Gentlemen...

116: Tank

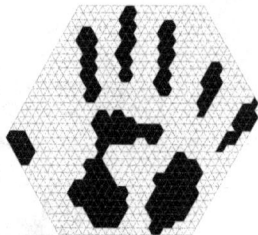

117: Hand Print

118: Exercise

119: Nemo

120: Sherlock Holmes

121: Earth

122: Flower

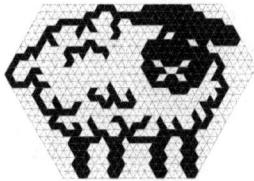

123: Hedgehog

124: Sheep

125: Bye Bye, Dribbel

126: Dan Fleming - Skunk

127: Weight Lifting

128: Megaphone

129: Skiing

130: Gun

131: Honey Comb   **132: Palm Tree**

133: Picasso - Flamingo

134: Wyoming

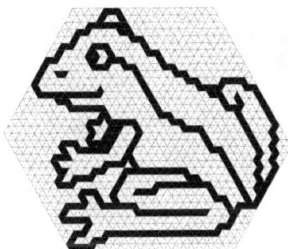

135: Frog

136: Hedgehog

137: Caparezza

138: Tango Silhouette

139: Glasses

140: Arctic Vehicle

141: Tiger

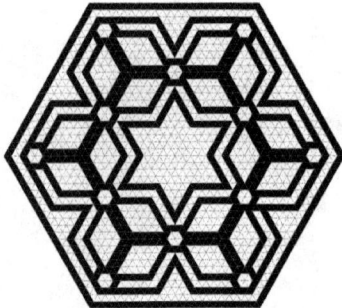

142: Star Design

143: Popeye

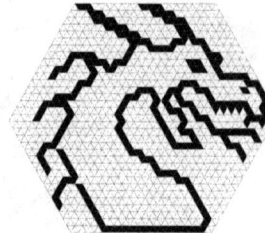

144: Dragon

145: Black Cat

146: Anchor

147: Penguin

148: King of the Jungle

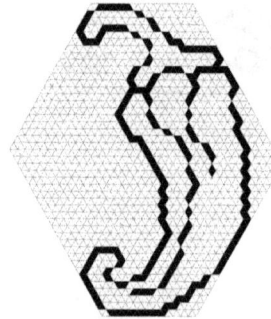

149: Pepper

150: Will you Marry Me?

151: Dolphin

152: Crossbow

153: Key

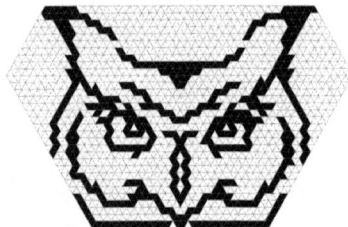

154: Owl

155: Spider

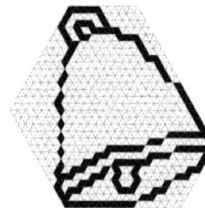

156: Bell

157: Bear

griddlers
Logic Puzzles

Picture Logic Puzzles:

Griddlers

Griddlers are picture logic puzzles in which cells in a grid have to be colored or left blank according to numbers given at the side of the grid to reveal a hidden picture.

Triddlers

Triddlers are logic puzzles, similar to Griddlers, with the same basic rules of solving. In Triddlers the clues encircle the entire grid. The direction of the clues is horizontal, vertical, or diagonal.

MultiGriddlers

MultiGriddlers are large puzzles that consist of several parts of common griddlers. A Multi can have 2 to 100 parts. The parts are bundled and, once completed, create a bigger picture.

Word Search Puzzles:

Word Search

Word Search is a word game that is letters of a word in a grid. The goal of the game is to find and mark all the words hidden inside the grid. The words may appear horizontally, vertically or diagonally, from top to bottom or bottom to top, from left to right or right to left. A list of the hidden words is provided.

Each puzzle has some text and underscores (_ _ _) to indicate missing word(s). If the puzzle was solved successfully, the remaining letters pop up in the grid and the missing words appear in the text.

Smart Things Begin With Griddlers.net

gRiddLeRs
Logic Puzzles

Number Logic Puzzles:

Sudoku

Sudoku is a logic-based, number-placement puzzle. The goal is to fill a grid with digits so that each column and each row contain the digits only once.

Irregular Blocks (Jigsaws)

Jigsaw puzzle is played the same as Sudoku, except that the grid has Irregular Blocks, also known as cages.

Killer Sudoku

The grid of the **Killer Sudoku** is covered by cages (groups of cells), marked with dotted outlines. Each cage encloses 2 or more cells. The top-left cell is labeled with a cage sum, which is the sum of all solution digits for the cells inside the cage.

Kakuro

Kakuro is played on a grid of filled and barred cells, "black" and "white" respectively. The grid is divided into "entries" (lines of white cells) by the black cells. The black cells contain a slash from upper-left to lower-right and a number in one or both halves. These numbers are called "clues".

Binary

Complete the grid with zeros (0's) and ones (1's) until there are just as many zeros and ones in every row and every column.

Smart Things Begin With Griddlers.net

gRiDDLeRS
Logic Puzzles

Number Logic Puzzles:

Greater Than / Less Than

Greater Than (or **Less Than**) Sudoku has no given clues (digits). Instead, there are "Greater Than" (>) or "Less Than" (<) signs between adjacent cells, which signify that the digit in one cell should be greater than or less than another.

Futoshiki

Futoshiki is played on a grid that may show some digits at the start. Additionally, there are "Greater Than" (>) or "Less Than" (<) signs between adjacent cells, which signify that the digit in one cell should be greater than or less than another.

Kalkudoku

The grid of the **Kalkudoku** is divided into heavily outlined cages (groups of cells). The numbers in the cells of each cage must produce a certain "target" number when combined using a specified mathematical operation (either addition, subtraction, multiplication or division).

Straights

Straights (**Str8ts**) is played on a grid that is partially divided by black cells into compartments. Compartments must contain a straight - a set of consecutive numbers - but in any order (for example: 2-1-3-4). There can also be white clues in black cells.

Skyscrapers

The **Skyscrapers** puzzle has numbers along the edge of the grid. Those numbers indicate the number of buildings which you would see from that direction if there was a series of skyscrapers with heights equal the entries in that row or column.

Smart Things Begin With Griddlers.net

www.ingramcontent.com/pod-product-compliance
Lightning Source LLC
Chambersburg PA
CBHW051338200326
41519CB00026B/7468